Studies in Computational Intelligence

Volume 658

Series editor

Janusz Kacprzyk, Polish Academy of Sciences, Warsaw, Poland
e-mail: kacprzyk@ibspan.waw.pl

About this Series

The series "Studies in Computational Intelligence" (SCI) publishes new developments and advances in the various areas of computational intelligence—quickly and with a high quality. The intent is to cover the theory, applications, and design methods of computational intelligence, as embedded in the fields of engineering, computer science, physics and life sciences, as well as the methodologies behind them. The series contains monographs, lecture notes and edited volumes in computational intelligence spanning the areas of neural networks, connectionist systems, genetic algorithms, evolutionary computation, artificial intelligence, cellular automata, self-organizing systems, soft computing, fuzzy systems, and hybrid intelligent systems. Of particular value to both the contributors and the readership are the short publication timeframe and the worldwide distribution, which enable both wide and rapid dissemination of research output.

More information about this series at http://www.springer.com/series/7092

Tomasz Pełech-Pilichowski
Maria Mach-Król · Celina M. Olszak
Editors

Advances in Business ICT: New Ideas from Ongoing Research

 Springer

Editors
Tomasz Pełech-Pilichowski
AGH University of Science and Technology
Kraków
Poland

Celina M. Olszak
University of Economics in Katowice
Katowice
Poland

Maria Mach-Król
University of Economics in Katowice
Katowice
Poland

ISSN 1860-949X   ISSN 1860-9503 (electronic)
Studies in Computational Intelligence
ISBN 978-3-319-83680-5   ISBN 978-3-319-47208-9 (eBook)
DOI 10.1007/978-3-319-47208-9

This Springer imprint is published by Springer Nature
The registered company is Springer International Publishing AG
The registered company address is: Gewerbestrasse 11, 6330 Cham, Switzerland

Contents

Verification of Temporal Knowledge Bases as an Important
Aspect of Knowledge Management Processes in Organization   1
Maria Mach-Król and Krzysztof Michalik

The Role of Simulation Performance in Software-in-the-Loop
Simulations   17
Tommy Baumann, Bernd Pfitzinger, Thomas Jestädt and Dragan Macos

Cognitum Ontorion: Knowledge Representation
and Reasoning System   27
Paweł Kaplanski and Pawel Weichbroth

Overview of Selected Business Process Semantization Techniques   45
Krzysztof Kluza, Grzegorz J. Nalepa, Mateusz Ślażyński, Krzysztof Kutt,
Edyta Kucharska, Krzysztof Kaczor and Adam Łuszpaj

Selected Approaches Towards Taxonomy of Business
Process Anomalies.   65
Anna Suchenia, Tomasz Potempa, Antoni Ligęza, Krystian Jobczyk
and Krzysztof Kluza

Hybrid Framework for Investment Project Portfolio Selection   87
Bogdan Rębiasz, Bartłomiej Gaweł and Iwona Skalna

Towards Predicting Stock Price Moves with Aid of Sentiment
Analysis of Twitter Social Network Data and Big Data
Processing Environment.   105
Andrzej Romanowski and Michał Skuza

On a Property of Phase Correlation and Possibilities to Reduce
the Walsh Function System   125
Lubomyr Petryshyn and Tomasz Pełech-Pilichowski

Introduction

The use of ICT has an increasing extent in business, the public sector and in private life. Opportunities offered by massive collection of data, as well as the Internet of Things, give a challenge for research institutes at universities and research and development departments in enterprises.

The issue of the effective use of ICT in the organization is not trivial. Advanced data processing algorithms are focused more on quantitative as well as qualitative data and text processing. In many cases, it is necessary to use artificial intelligence paradigms. Increasingly, there are also used data mining techniques, as well as advanced processing algorithms for signal diagnostic. In addition, the current high computing powers, including the ability to use cloud computing technology, allow the use of advanced algorithms, which until now are far too much time consuming. Moreover, organizations often use a variety of ICT systems not fully cooperating with each other. The necessity of extraction from multiple data sources of various formats and with varying degrees of coherence is also identified. In addition, the inability to create or maintain business teams of qualified data analysts, who would allow for the construction of their own, dedicated solutions, and at the same time the use of the current possibilities of science, was observed. These prerequisites are not only a necessity but also a motivation for undertaking research in the area of decision support, business intelligence, data mining, and big data.

The book contains papers, which are the result of active discussions held at the international conference Advances in Business ICT (ABICT) in the years 2014 (Warsaw, Poland) and 2015 (Lodz, Poland). ABICT provides an international, multidisciplinary forum for scientists and experts from research units at universities and in industry, which allows to explore new ideas, approaches, research directions, tools, and verification of applied business intelligence solutions.

Researchers, data analysts, entrepreneurs and IT professionals will find this book interesting.

Tomasz Pełech-Pilichowski
Maria Mach-Król
Celina M. Olszak

Verification of Temporal Knowledge Bases as an Important Aspect of Knowledge Management Processes in Organization

Maria Mach-Król and Krzysztof Michalik

Abstract The paper deals with the problem of temporal knowledge verification treated as an important process of knowledge management (KM). At the same time authors postulate addition of knowledge verification (KV) to the set of several recognized processes in the theory of V&V. Moreover the motivation for implementing a temporal knowledge base system is presented, the implementation methodology is outlined, and the KV process is described in detail, using the example of the Logos semantic reasoner. The main achievements of the paper are: elaborating a new implementation methodology for a temporal knowledge base system, and elaborating detailed KV steps as well as viewing the KV as a process in KM.

Keywords Organizational creativity · Temporal knowledge base system · Verification · Implementation methodology · Knowledge management · Logos semantic reasoner

1 Introduction

Organizational creativity is a relatively new concept in the theory of management, which has partially arisen on the ground of knowledge management (KM).

There are many definitions of organizational creativity, but it is commonly perceived as a team, dynamic activity, responding to changing features of organization's environment, a team process—see e.g. [4, 28].

The organizational creativity is therefore to be perceived in the context of organizational dynamics, because it depends on the situational changes and is composed of processes. Therefore while discussing the question of computer support for organizational creativity, the temporal aspects should not be omitted.

M. Mach-Król (✉) · K. Michalik
University of Economics, Bogucicka 3, 40-226 Katowice, Poland
e-mail: maria.mach-krol@ue.katowice.pl

K. Michalik
e-mail: krzysztof.michalik@ue.katowice.pl

© Springer International Publishing AG 2017
T. Pełech-Pilichowski et al. (eds.), *Advances in Business ICT: New Ideas from Ongoing Research*, Studies in Computational Intelligence 658,
DOI 10.1007/978-3-319-47208-9_1

Such a way of formulating this problem—underlining its dynamic aspect—justifies a proposal of using an intelligent system with a temporal knowledge base, as a tool supporting creation and development of organizational creativity, which is understood as organizational asset (see e.g. [14, 27]).

By the system with a temporal knowledge base we will understand (slightly modifying the definition given in [15]) an artificial intelligence system, which explicitly performs temporal reasoning. Such a system contains not only fact base, rule base, and inference engine, but also directly addresses the question of time. For an intelligent system to be temporal, it should contain explicit time representations in its knowledge base—formalized by the means of temporal logics—and at least in the representation and reasoning layers. In the paper we use an example of the Logos tool—a reasoning system constructed by the authors within the frame of the research project under the same name. One of the assumptions of the Logos project is the possibility of using it for different scientific researches and experiments, among others using it for building temporal knowledge base research prototype.

The main aim of the paper is to discuss the current and predicted knowledge anomalies that have to be detected and successfully resolved by the system with a temporal knowledge base in order to manage knowledge in organization. The second aim is to present a new implementation methodology for a temporal knowledge base system supporting organizational creativity, and to present the steps performed during validation and verification of the temporal knowledge embedded in the system.

2 Motivation

While discussing the use of any computer tool, first of all one has to take into account the features of the domain to be supported. This applies also to systems with a temporal knowledge base and their application in supporting organizational creativity.

Some elements that justify the use of an intelligent tool with direct time references, may be found in the definitions of organizational creativity:

- [26, 31] claim that the effects of organizational creativity encompass ideas and processes—which in our opinion should be referred to as creative knowledge. The knowledge is to be codified and stored in a knowledge base, and because it is a changing knowledge, the knowledge should be a temporal one;
- in the definition given by [3] the author points out that organizational creativity is more heuristic than algorithmic in nature (p. 33)—therefore it is not possible to use classical analytical tools, because heuristic tasks lack of algorithmic structure, they are complex and uncertain (see e.g. [1] p. 6);
- [28] suggests that ideas born during creative processes (that is, the creative knowledge) must be adequate to the situation (p. 289). Therefore they have to change dynamically, because the situation of organization also constantly changes;

- the changeability, dynamics, and process nature of organizational creativity, which justify its codification in a temporal knowledge base, are stressed in definitions given by [2, 5, 6, 10, 17];
- [4] point out that organizational creativity must be analyzed on individual, group, and organizational levels. This justifies the use of a knowledge base: if the creativity (its effects) is to penetrate between the levels, to support collaboration, a system with a temporal knowledge base enables such penetration;
- the justification for using temporal formalisms to codify creative knowledge may be found in the definitions given by [22, 23] where authors point out the badly structured nature of creative problems. One of temporal formalisms' advantages is the possibility to formalize unstructured problems.

While reading many authors' discussions on the essence of organizational creativity, one sees that this is primarily team activity. The effect of this activity may be referred to as "creative knowledge", which itself generates new ideas, concepts, and solutions. To do so, the creative knowledge must be first codified, and next disseminated. This justifies the use of a knowledge base system. But the creative knowledge changes in time, due to several reasons.

First, organizational creativity is a process, therefore its effects are subject to change. Moreover, the process encompasses solving problems that also change, because the organization's environment changes [15] pp. 13–15, [8] p. 150 and next, 176 and next.

Second, each knowledge—including the creative one—changes simply with the passage of time, with the flow of new information about objects [7].

Third, organizational creativity is linked with dynamics, which can be seen e.g. in the assets approach to this creativity or in the requirement of adapting creative knowledge to situational context.

All the above leads to conclusion that a knowledge base system is not enough to support organizational creativity, because classical knowledge bases do not support time. Therefore in this paper we propose the use of a temporal knowledge base system, as defined earlier. Such system is able to perform the tasks arising from the characteristics of organizational creativity and its artifacts.

An important element of the implementation methodology of the proposed system is the process of validation and verification of temporal knowledge embedded in the system. The system is a rule-based one (strictly speaking: a temporal rule-based one), thus in order to run properly, its knowledge base must be correct. And a "correct" temporal knowledge base means that it has been verified and validated to ensure that there are no anomalies such as:

- redundant rules,
- subsuming rules,
- contradictory rules,
- unused attributes,
- unused values,
- recursive rules (inference loop, circularity).

Thus, the procedures of validation and verification of knowledge (KV) constitute an important fragment of implementation methodology for the proposed system.

3 Related Work

In the literature, there are many methodologies concerning knowledge-based systems, among others [19], p. 136:

- blackboard architecture,
- KADS and CommonKADS,
- HyM for hybrid systems,
- Protégé,
- CAKE.

It must be however noted, that the above mentioned methodologies have been created for expert systems, while the proposed temporal knowledge base system is not a typical ES. In the context of its architecture, it is worthy of considering the blackboard architecture, which is by some authors understood as a knowledge engineering methodology [12]. It enables to explicitly represent knowledge and its structure in a rule-based system (and the proposed temporal knowledge base system is a rule-based one). It may be acknowledged that a postulated division of system's knowledge base into several sub-bases means implementing the blackboard architecture and achieving its assumptions. The advantages of such an approach are as follows[1]:

- possibility of using the creative knowledge from many participants of the creative process,
- group working, and brainstorming support,
- easy implementation of creative knowledge chunks in a formalized manner,
- possibility of differentiating knowledge representation forms, and ways of reasoning (that is, possibility of using more than one temporal formalism),
- facilitation of processing the creative knowledge coming from heterogeneous sources—that is knowledge created in various ways, by different means, and during different stages of organizational creativity process. In the case of heterogeneous knowledge sources it should be added they may require also hybrid architecture, as in the case of hybrid expert system shell PC-Shell [18].

The second interesting methodology is CAKE (Computer Aided Knowledge Engineering), elaborated by Michalik. The detailed description of CAKE may be found e.g. in [19, 21]. Its advantages are similar to those of the blackboard methodology:

- use of the blackboard systems methodology (with all its advantages),

---

[1]Based on [19], p. 139.

- easy management of heterogeneous knowledge sources,
- support of group working,
- automatic control of formalized creative knowledge code,
- knowledge base editor,
- a package of wizards facilitating the coding process of the acquired knowledge.

It has to be pointed out, however, that the knowledge coding formalism, embedded in the CAKE system, has no temporal references. A sample diagram of knowledge base anomalies can be found e.g. in [29] and more detailed discussion on verification and validation in [24]. Some introductory concepts concerning KV in Logos system being subject of our presentation can be found in [20]. Very interesting remarks on KV in the context of knowledge engineering in the CommonKADS methodology can be found in [25]. Authors differentiate between internal validation for both internal and external meaning, e.g. saying that some people use the term verification for internal validation and apply validation concept against user requirements ("is it the model right?").

4   System Implementation Methodology

It is also important that the aforementioned methodologies relate mainly to building expert systems, while the methodology needed for a temporal knowledge base system has to take into account also the processes of implementing the system in a creative organization. Therefore it is not possible to directly use any of the aforementioned methodologies, and a new one has to be developed, suited to the task of supporting organizational creativity by a temporal knowledge base system. Two very important questions thus arise.

First, the proposed system is an intelligent one, containing at least one temporal knowledge base, therefore the implementation methodology has to make use of (but not copying directly) existing methodologies for implementing such systems, as e.g. expert ones.

Second, the main aim of the system is to support organizational creativity, therefore the most important system elements are user interface and knowledge base. The first enables both adding creative knowledge to the system, and querying this kind of knowledge, the second is needed in representation and reasoning layers. The proposed methodology should accommodate also these elements.

The implementation methodology for a temporal knowledge base system has to be conformable to temporal knowledge base system's lifecycle. We propose the following lifecycle for the system (adapted from [11]):

1. Problem identification, and definition of users' needs;
2. System's formal specification, encompassing dialogues with users;
3. Definition of knowledge sub-bases', and general knowledge base structure and scope, choice of knowledge representation technique(s), creation of reasoning algorithm;

4. Creative knowledge acquisition;
5. Prototype creation and verification;
6. System coding and testing;
7. System maintaining and development—principally the creative knowledge bases and user interface.

The implementation methodology for the temporal knowledge base system covers points 1, 2, 4–6, and 7 of the proposed system's lifecycle.

The methodology should be focused primarily on the creative knowledge (its continuous acquisition and representation), and on user interface. Therefore its main elements are creative knowledge engineering, and system engineering, with emphasis put on interface design and prototyping. The general structure of the proposed methodology is presented in Fig. 1.

The proposed methodology has been inspired by other knowledge engineering methodologies for the process of knowledge management—particularly by the work [32]—and by classical, fundamental methodologies for implementing expert systems: [30], pp. 135–139, and [9], p. 139. Obviously, it was not possible to directly merge the existing models, the proposal concerning knowledge engineering had to be remodeled in the context of organizational creativity process, while methodologies for implementing expert systems had to be adapted to the temporal knowledge base system, and its main task.

The methodology for implementing a temporal knowledge base system, presented in Fig. 1, starts with the group of activities concerning capturing, and modeling of creative knowledge. At this stage it is essential to discover creative processes running within the team of employees involved in organizational creativity. This will enable to identify needs concerning the creative knowledge, and its usage by an organization (or team). Having this information, the next step of the methodology is to choose and/or design tacit creative knowledge acquisition methods, as well as to acquire explicit creative knowledge. This is so because we assume that the creative knowledge, as any other kind of knowledge, may be divided into tacit and explicit one. Next, it is necessary to identify (with the aid of previously gathered information) tacit knowledge, and sources of explicit knowledge, and to acquire both types of creative knowledge. Only then it is possible to model and analyze the creative knowledge, which is to be incorporated in the temporal knowledge base system.

During each stage of the proposed methodology, especially during the creative knowledge engineering stage, it is indispensable to closely cooperate with system users, that is the employees involved in the process of organizational creativity. Without them it is impossible to identify, and to acquire tacit knowledge. Moreover, the system will be useful only if people want to use it.

Activities concerning creative knowledge modeling, implementation, and verification are absolutely crucial, therefore in the proposed methodology there is a possibility to return to previous stages, in order to refine knowledge representation and implementation, or even to completely change the design of the knowledge model.

It also has to be explained why activities concerning system's specification, design, and implementation are placed at the end of the methodology, which differs from

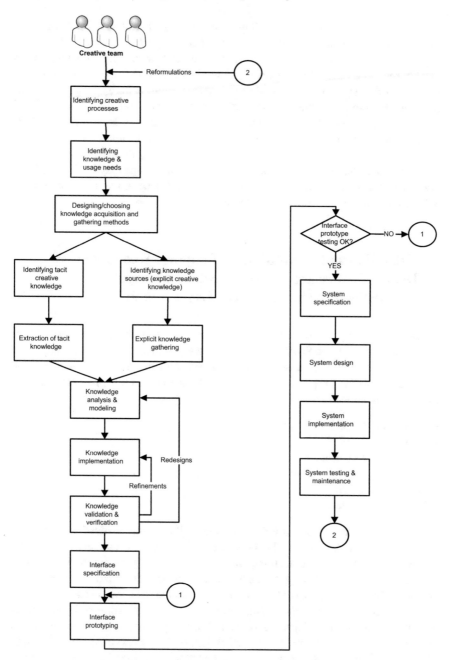

Fig. 1 Schema of temporal knowledge base system implementation methodology

classical implementation methodologies for intelligent systems. As it has been already said, the main task for the temporal knowledge base system is to support organizational creativity, so its most essential elements are temporal creative knowledge base(s) and GUI. Thus the methodology is focused on these elements. Activities concerning system's engineering are also important, but are of ancillary nature regarding temporal KB, creative knowledge management, and GUI design.

As it has been already said, an important element (step) in the proposed methodology concerns temporal knowledge validation and verification, to ensure that the knowledge base is correct. The V&V step is preceded by knowledge analysis and modeling, and knowledge implementation. These three steps may be followed several times, continuously refining the knowledge base. Due to the importance of the V&V procedure, it is described in detail in the next section

5   Validation and Verification of Temporal Knowledge During the Process of Implementing a Temporal Knowledge Base System—The Logos Example

According to Fig. 1 we regard verification as a part of temporal knowledge base implementation methodology. What's more, we postulate KV to be an important process in set of KM processes, widely accepted in the theory of management. Taking into account the process approach to KM, in the literature processes the most frequently mentioned are:

- knowledge acquisition,
- knowledge creation,
- knowledge gathering,
- knowledge sharing/dissemination,
- knowledge transfer,
- knowledge utilization.

It can be seen the lack of process concerning validation and verification of acquired or created knowledge. So all the subsequent processes can use incorrect knowledge with possibly important consequences for the given organization. For example process of knowledge sharing means also e-learning that can result in teaching bad practices, theory, regulations. The process of knowledge utilization often means making decisions or supporting them. This serious process of decision-making should be strongly based on correct knowledge. Therefore KM processes have to be supported by proper procedures of knowledge verification. In the context of automatic knowledge processing using AI methods it may concern knowledge bases, in context of our researches—temporal knowledge bases (see e.g. [16]).

Taking into account the above assumption, the verification subsystem is one of the intensively developed part of the presented Logos semantic reasoner. The fact that the inferential knowledge is basically rule-based part of its knowledge base, it can result in some logical incorrectenesses. We additionally assumed that temporal knowledge, in Logos also in the form of inference rules and facts, can include some specific logical errors. The third assumption is based on [13] and the fact that both temporal calculus of events as well as situation calculus can be formalized by means of Horn clauses, additionally augmented with negation by failure. It's the reason for which they can be also executed as a logic programs.

As mentioned earlier, we build the Logos semantic reasoner, being research software platform for our experiments with temporal knowledge bases as well as with temporal reasoning. Additionally, while developing this system, we take into account the mentioned issue of KV. At present main procedures discovering some anomalies in knowledge bases are ready. What has to be done—according to our thesis concerning KV in temporal knowledge bases—is the identification of specific, temporal anomalies and the efficient algorithms to detect them. The thesis is that in temporal knowledge bases some new, very specific anomalies and errors, may theoretically appear. On the other hand, most of KV methods already built-in in Logos for conventional (not temporal) knowledge bases are also useful and even necessary. The reason is that temporal knowledge bases may include the same kind of anomalies as the conventional ones. Most of the temporal anomalies we plan to detect as a first step while building Logos relate to incorrect time dependencies as declared in knowledge base. As we mentioned in Sect. 1, the correct temporal knowledge base means that it has been verified and validated to ensure that there are no anomalies such as the following [19, 21].

Redundancy
Two rules we regard as redundant, if for two rules:

$R_i \leftarrow W_{i1 \wedge \ldots \wedge} W_{in}$ and
$R_j \leftarrow W_{j1 \wedge \ldots \wedge} W_{jn}$, where $i \neq j$,
holds: $\{W_{i1} .. W_{in}\} = \{W_{j1} .. W_{jn}\}$

Subsuming Rules
If for two different rules:

$R_i \leftarrow W_{i1 \wedge \ldots \wedge} W_{im}$ and
$R_j \leftarrow W_{i1 \wedge \ldots \wedge} W_{in}$, where $i \neq j$,
holds $\{W_{i1}, .., W_{im}\} \subseteq \{W_{i1}, .., W_{in}\}$, then we say, that rule R_j subsumes R_j.

Contradictory Rules
Two rules we regard as contradictory if

$R_i \leftarrow W_{1 \wedge \ldots \wedge} W_n$ and
$\neg R_j \leftarrow W_{1 \wedge \ldots \wedge} W_n$, where $i \neq j$.

Inconsistent Rules
Two rules we regards as inconsistent if

$R_i \leftarrow W_{1 \wedge \ldots \wedge} W_n$ and
$R_j \leftarrow W_{1 \wedge \ldots \wedge} W_n$, where $i \neq j$ and $R_i \neq R_j$,

Incompleteness

We assumed that temporal knowledge base is complete when contains all possible combinations of attributes and their allowable values in rules antecedents and consequents. It should be noticed that in practice not all combinations are required, so this kind of verification is only warning for knowledge engineer that some rules could be missing.

Missing Rules

The anomaly called here as missing rules can be treated as a special case of incompleteness. While creating KV module of Logos system we consider missing rules as the case when some of decision-making attributes are not present in any of the rule antecedents. This situation can appear as side effect of rapid prototyping and incremental method of knowledge base development.

Unused Attributes and Values

Our system in order to be able to detect some anomalies and errors requires explicit declaration of attributes and values being used in the knowledge base. When given attribute or value is never used in any of rules then Logos gives warning addressed to knowledge engineer because it may be information about serious anomaly in knowledge base. On the other hand, similarly as in the case of missing rules it can be side effect of using methodology of rapid prototyping and incremental development of knowledge base.

Recursive Rules (Inference Loop, Circularity)

Recursion in most of rule-based systems is very important anomaly in knowledge base with serious consequences and sometimes very difficult to detect by knowledge engineer without software support, e.g. as implemented in our Logos system. Recursion in this context may take a variety of patterns. In the simplest case can be like this:

$$K_i \leftarrow W_1 \text{ and } K_i = W_1.$$

This type of recursion (loop) is very easy to detect for knowledge engineer, even without special algorithms. The sign '=' does not mean simple equality but may have more complex semantics of ability of two expressions to match. Other variants of the same direct recursive call of the conditions to the conclusions of the rule may take one of the following schemes:

$$K_i \leftarrow W_1 {}_\wedge.. {}_\wedge W_n \text{ and } K_i = W_1,$$
$$K_i \leftarrow W_1 {}_\wedge.. {}_\wedge W_n \text{ and } K_i = W_n,$$
$$K_i \leftarrow W_1 {}_\wedge..W_j.. {}_\wedge W_n \text{ and } K_i = W_j,$$

Where '=' means rather ability to match or unify than identity or equality.

While recursion in rules of logic programs is very useful and correct situation (provided their correct semantics), the recursion in expert systems is generally treated (as mentioned) as serious knowledge base anomaly. For example, the correct recursion in a simple logic program, not an infinite loop (one rule and one fact):

p([]).
p([X|Y]) ← p(Y).

We rejected using recursions in Logos for many reasons, the main is that Prolog is first of all programming formalism and not knowledge representation language for temporal knowledge bases.

Much more difficult situation to detect by the knowledge engineer is that of indirect recursion. Sometimes, in large knowledge bases with several levels of inference, it can be practically not possible to detect in reasonable time. Then software support as e.g. that we implemented in Logos is absolutely necessary. In such case of indirect recursive call it does not occur at the same level of a given rule to its conclusion, within a single rule. This can be illustrated by the following example (see Fig. 2):

$R_1: K_1 \leftarrow W_{11} \wedge .. W_{1j} .. \wedge W_{1x}$
$R_i: K_i \leftarrow W_{i1} \wedge .. W_{il} .. \wedge W_{iy}$
$R_n: K_n \leftarrow W_{n1} \wedge .. W_{nm} .. \wedge W_{nz}$
where: $W_{1j} = K_i i \ W_{il} = K_n . i \ W_{nm} = K_1$ and '=' means matching/unification.

In this case, the recursive call is related to a lower level in the hierarchy of rules, making its location is difficult to detect for a knowledge engineer. NB: Incidentally, defined earlier contradiction of rules—as mentioned—can also be caused by indirect inference.

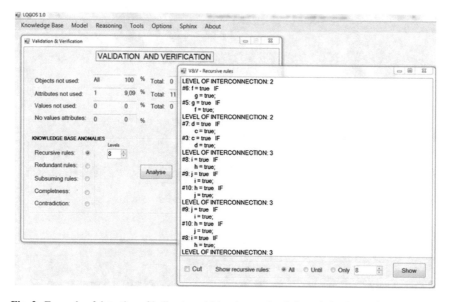

Fig. 2 Example of detection of indirect, multi-level recursion in knowledge base of Logos

Direct contradiction:

$p \leftarrow q$
$\neg p \leftarrow q.$

Indirect contradiction appearing during inference process:

$\neg p \leftarrow r$
$p \leftarrow q$
$q \leftarrow r.$

6 Explanations

The logical analysis executed by KV subsystem can be not sufficient to evaluate valid-
ity of temporal knowledge base and all aspects of correctness. To support semantic
level of knowledge base analysis and to add the facility for knowledge engineer
or end-user we built wide spectrum of explanations. Their names are taken from
terminology of knowledge-based systems, in that the three built earlier by authors
(PC-Expert (1986–1987), Diagnostician (1988), PC-Shell (1990–2006). In current
version of Logos the following explanation facilities are available (see Fig. 3):

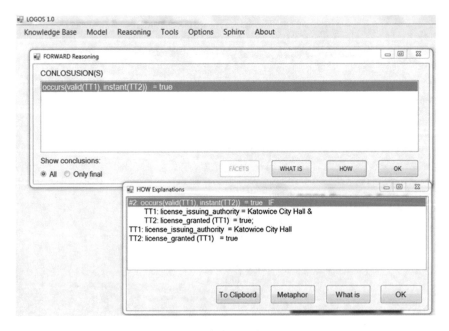

Fig. 3 Example of solution and explanation facilities of Logos

- *How*—which shows how the selected solution has been logically derived. *How* explanations (if the blackboard architecture is being used) shows also how the fact has been derived during consultation.
- *Why*—this kind of explanation is the answer to the system' question, why some query has been generated while reasoning. On the one hand it can be treated as a reason to ask the user the specific question, on the other hand it should be considered as an important element of verification and location of errors in the knowledge base, both at the logical and semantic level.
- *What is*—explanations, in fact—being also additional support for How and Why verification role—provide more detailed, textual information also about conclusions, as well as some questions. It is also possible to attach some explanations to facts, in the form of *facts descriptions*, e.g.: source of information or availability of deeper/further explanations.
- *Metaphors*—enable knowledge engineer to attach more textual information about selected rules, what can be useful at initial stage of using the application but also to check semantic correctness of the underlying rules by knowledge engineer.
- *Facts descriptions*—It is also possible to attach some additional explanations to facts, e.g.: source of knowledge or availability of deeper/further explanations, what is also of great importance in some circumstances for knowledge engineer.

7   Concluding Remarks

Our main goal at this stage of research was to design and to develop a software platform to make experiments with temporal knowledge bases. In this paper we focused on the problem on KV, which is always present while building knowledge bases, but sometimes underestimated. It's also part of the suggested (see Fig. 1) methodology The consequences of badly evaluated knowledge bases in this respect can be very serious, and for example some anomalies can't be detected for long time resulting in bad decisions. The more so, as we have shown, some of the anomalies, as for example recursions (inference loops) can be sometimes very difficult to detect by knowledge engineer. So our thesis since the beginning of the project was to build computer aided knowledge engineering system automatically supporting knowledge engineer, especially in such difficult to analyze cases. The next step in our researches concerning temporal knowledge bases will be identification of the specialized anomalies typical only for temporal systems. We suspect that beside the simple errors related to time we may discover special kind of temporal anomalies. Even some of already identified and described in our paper some anomalies can take the new character in relation to time. We have also additional hypothesis that temporal knowledge bases have kind of anomalies which are completely specific and different from the described ones. Verification of these hypothesis we take as next goal of our researches.

References

1. Aggarwal, A., 2001. A Taxonomy of Sequential Decision Support Systems. Kraków, pp. 1–11.
2. Alvarado, L. D., 2006. The creative organizations as living systems. In: S. Torre and V. Violant, Eds. *Understanding and evaluating creativity*. Malaga: Editiones Algiba, pp. 375–382.
3. Amabile, T. M., 1996. Creativity in Context: Update to The Social Psychology of Creativity. Boulder: Westview Press.
4. Andriopoulos, C. and Dawson, P., 2014. *Managing Change, Creativity and Innovation. Second Edition*. Los Angeles/London/New Delhi/Singapore/Washington DC: SAGE Publications.
5. Baer, M., 2012. Putting Creativity to Work: The Implementation of Creative Ideas in Organizations. *Academy of Management Journal*, 1 October, 55(1), pp. 1102–1119 .
6. Basadur, M., Basadur, T. and Licina, G., 2012. Organizational Development. In: M. D. Mumford, Ed. *Handbook of Organizational Creativity*. London/Waltham/San Diego: Elsevier Inc., pp. 667–703.
7. Benthem van, J., 1995. Temporal Logic. In: D. M. Gabbay, C. J. Hogger and J. A. Robinson, Eds. *Handbook of Logic in Artificial Intelligence and Logic Programming. Volume 4: Epistemic and Temporal Reasoning*. Oxford: Clarendon Press, pp. 241–350.
8. Czaja, S., 2011. Czas w ekonomii. Sposoby interpretacji czasu w teorii ekonomii i w praktyce gospodarczej. Wrocław: Wydawnictwo Uniwersytetu Ekonomicznego.
9. Hayes-Roth, F., Waterman, D. and Lenat, D. Eds., 1983. *Building Expert Systems*. Reading, Mass.: Addison-Wesley Publishing Company.
10. Hirst, G., Knippenberg, D. v. and Zhou, J., 2009. A Cross-Level Perspective on Employee Creativity: Goal Orientation, Team Learning Behavior, and Individual Creativity. *Academy of Management Journal*, 1 April, 52(2), pp. 280–293.
11. Infernetica, 2012. *Systemy ekspertowe dla biznesu*. [Online] Available at: http://infernetica.com/systemy-ekspertowe/ [Accessed: 07 04 2014].
12. Kendal, S. and Creen, M., 2007. *An Introduction to Knowledge Engineering*. London: Springer.
13. Kowalski R, Sergot M., 1986. A logic-based calculus of events. New Generation Computing, March, vol. 4, pp 67–95.
14. Krupski, R., Ed., 2011. *Rozwój szkoły zasobowej zarządzania strategicznego*. Wałbrzych: Wałbrz. Wyż.Szk.Zarz. i Przedsibę.
15. Mach, M. A., 2007. Temporalna analiza otoczenia przedsiębiorstwa. Techniki i narzędzia inteligentne. Wrocław: Wydawnictwo AE.
16. Mach-Król M., Michalik K., 2014. *Selected aspects of temporal knowledge engineering*. In: Ganzha, M., Maciaszek, L., Paprzycki, M. (Eds.), Proceedings of the 2014 Federated Conference on Computer Science and Information Systems. Annals of Computer Science and Information Systems, Volume 2. ISSN 2300-5963, ISBN 978-83-60810-58-3 (Web), 978-83-60810-57-6 (USB), 978-83-60810-61-3 (ART) IEEE Catalog Number: CFP1485N-ART (ART), CFP1485N-USB (USB). doi: 10.15439/978-83-60810-58-3
17. Martins, E. C. and Terblanche, F., 2003. Building organisational culture that stimulates creativity and innovation. *European Journal of Innovation Management*, February, 6(1), pp. 64–74.
18. Michalik K., 2010, PC-Shell/SPHINX jako narzędzie tworzenia systemów ekspertowych. In: J. Gołuchowski and B. Filipczyk (eds.), Systemy ekspertowe – wczoraj, dziś i jutro; wiedza i komunikacja w innowacyjnych organizacjach, Katowice: Wydawnictwo UE.
19. Michalik, K., 2014. Systemy ekspertowe we wspomaganiu procesów zarządzania wiedzą w organizacji. Katowice: Wydawnictwo Uniwersytetu Ekonomicznego.
20. Michalik, K., 2015. Validation and Verification of Knowledge Bases in the Context of Knowledge Management. Logos Reasoning System Case Study, [in:] Technologie wiedzy w zarządzaniu publicznym'13, ed. J. Gołuchowski, A. Frączkiewicz-Wronka, Wydawnictwo UE, Katowice.
21. Michalik, K., Kwiatkowska, M. and Kielan, K., 2013. Application of Knowledge-Engineering Methods in Medical Knowledge Management. In: R. Seising and M. E. Tabacchi, Eds. *Fuzziness and Medicine: Philosophical Reflections and Application Systems in Health Care*. Berlin Heidelberg: Springer, pp. 205–214.

22. Mumford, M. D., Medeiros, K. E. and Partlow, P. J., 2012. Creative Thinking: Processes, Strategies, and Knowledge. *The Journal of Creative Behavior*, March, 46(1), pp. 30–47.
23. Mumford, M. D., Robledo, I. C. and Hester, K. S., 2011. Creativity, Innovation and Leadership: Models and Findings. In: A. Bryman, et al., Eds. *The SAGE Handbook of Leadership*. London: SAGE Publications Ltd., pp. 405–421.
24. Owoc M., Ochmańska M., Gładysz T., 1999. On Principles of Knowledge Validation, [in:] Validation and Verification of Knowledge Based Systems: Theory, Tools and Practice, eds. A. Vermessan, F. Coenen, Kluwer Academic Publishers, Boston.
25. Schreiber et al., 2000. Knowledge Engineering and Management, The CommonKADS Methodology, The MIT Press, Cambridge MA.
26. Shalley, C. E., Gilson, L. L. and Blum, T. C., 2000. Matching Creativity Requirements and the Work Environment: Effects on Satisfaction and Intentions to Leave. *Academy of Management Journal*, 1 April, 43(2), pp. 215–223.
27. Sirmon, D. G., Hitt, M. A., Ireland, R. D. and Gilbert, B. A., 2011. Resource Orchestration to Create Competitive Advantage: Breadth, Depth, and Life Cycle Effects. *Journal of Management*, September, Vol. 37(No. 5), pp. 1390–1412.
28. Unsworth, K. L., 2001. Unpacking Creativity. *Academy of Management Review*, Vol. 26(No. 2), pp. 286–297.
29. Vermessan A.I., 1998. Foundation and Application of Expert System Verification and Validation, [in:] The Handbook of Applied Expert Systems, ed. J. Liebowitz, CRC Press, New York 1998.
30. Waterman, D., 1986. *A Guide to Expert Systems*. Reading, Mass.: Addison-Wesley Publishing Company.
31. Woodman, R. W., Sawyer, J. E. and Griffin, R. W., 1993. Toward a Theory of Organizational Creativity. *The Academy of Management Review*, April, Vol. 18(No. 2), pp. 293–321.
32. Yazdanpanah, A. and Sadri, R., 2010. Proposed Model for Implementation Expert System for the Planning of Strategic Construction Projects as a Tool for Knowledge Management. Istanbul, IPMA.

The Role of Simulation Performance in Software-in-the-Loop Simulations

Tommy Baumann, Bernd Pfitzinger, Thomas Jestädt
and Dragan Macos

Abstract The simulation performance is one aspect that becomes important in the practical application of simulations. A number of situations necessitates high-performance simulations: The simulation of large systems, over long time-periods or exploring a large solution space. Starting with these scenarios we discuss the parallelization of a discrete event simulation (DES) model using a synthetic benchmark model. Using the example of the German automatic toll system we explore the performance constraints originating from coupling an abstract simulation model with a real-world system.

1 Introduction

Simulations—regardless of the chosen technology—are an integral part of the system development process. The ever increasing complexity of software-intensive systems puts a particular strain on the technical support for automating the system development [1]—requirements could be undocumented, written in natural language, missing the user or operational perspective [2, 3] and of course incorrectly implemented.

T. Baumann (✉)
Andato GmbH & Co. KG, Ehrenbergstraße 11, 98693 Ilmenau, Germany
e-mail: tommy.baumann@andato.com

B. Pfitzinger · T. Jestädt
Toll Collect GmbH, Linkstraße 4, 10785 Berlin, Germany
e-mail: bernd.pfitzinger@toll-collect.de

T. Jestädt
e-mail: thomas.jestaedt@toll-collect.de

D. Macos
Beuth Hochschule Für Technik Berlin, Luxemburger Str. 10, 13353 Berlin, Germany
e-mail: dmacos@beuth-hochschule.de

© Springer International Publishing AG 2017
T. Pełech-Pilichowski et al. (eds.), *Advances in Business ICT: New Ideas from Ongoing Research*, Studies in Computational Intelligence 658,
DOI 10.1007/978-3-319-47208-9_2

Accordingly many efforts exist to improve the development process—test driven development [4] starts with the implementation of the (unit) test cases that need to be passed by the system, agile process models [5] reduce the development risk by shortening the development period.

However, both approaches fall short of putting the development effort into the operational context: The success of a system is only determined once it is used under normal operating conditions. I.e. the usage, operational capacities and practices as well as the interplay with neighboring systems determines whether a system fulfills its requirements. Merging development and operations ("DevOps", [6]) is one proposed solution—albeit increasing the operational risks to a level unacceptable in liability-critical systems or when a system needs to be non-discriminatory. In contrast, 'micro-services' split a complex system into simple, seemingly independent services and claim to eliminate the complexity of large systems [7].

The use of simulation models is a different approach to overcome these problems: A simulation model is put at the beginning of the system development—capturing the operational context, the current knowledge regarding the requirements and encompasses the whole system (including user behavior, operational procedures and partner systems). As in "test-driven", the simulation model defines the test case and acceptance criteria—at the level of the complete, integrated system. In that sense, requirements become *executable* through models before the system development starts. During the system development effort, requirements are first included in the simulation model before the implementation starts (see Fig. 1).

Even when the simulation model remains at a certain level of abstraction (i.e. not connected to the real-world system) the simulation performance becomes a bottleneck: To realistically reproduce the operational context simulation parameters need to

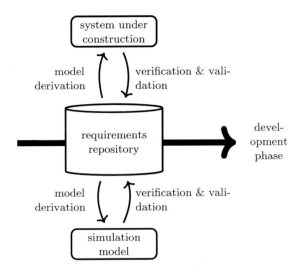

Fig. 1 In the simulation driven development approach (SDD) the requirements become *executable* as a simulation model

be fitted to observations (e.g. see [8] for the German automatic toll system). The simulation performance is even more important when the simulation model is intended to interface with real-world (sub-) systems running in 'real time'—simulation models need to be parallelized (see [9] where we introduced a simple benchmark model).

The ability to simulate at different levels of abstraction is of particular interest when simulations accompany or drive the system development process. To that extent the next section introduces the coupling between an abstract simulation model and a real-world system—either as a hardware or software-intensive system. We describe a particular challenge: The time synchronization across the whole system when the level of abstraction changes during a simulation run, e.g. a real-world system replaces a simulated sub-system starting at a given point in time. Retaining a valid concept of time spanning the simulated and the real-world system is covered in more detail in Sect. 3. In particular the section summarizes the experiences gathered from a synthetic benchmark model mimicking the interplay of a simulation model exchanging data with a web-server using thousands of parallel TCP/IP connections simultaneously. Section 4 takes the example of the German automatic toll system as a potential use case. We estimate the performance constraints for the simulation by observing the dynamics of this real-world system.

2 Hardware and Software in the Loop Simulation

Characteristic to a simulation driven design approach is the integrated system level development scope accompanying all design stages. Over time the simulation model of the system under development is refined continuously as the development progresses from the requirement phase up to the implementation phase. The requirements repository (see Fig. 1) collects the current knowledge regarding the system under construction which is in turn first implemented as an executable simulation model—putting any requirement immediately into the context of the integrated system under normal operating conditions. Operational scenarios are either set up in the requirements phase or exist from the operational experience gathered in an existing system.

With the progress of the development process already existing or newly implemented sub-systems may be tested against the holistic system model (including all sub-systems) for validation, optimization and to locate system-level integration issues. To that extend an abstract simulation model may be used to produce dynamic performance indicators for sub-systems—to be tested independently. More stringent is the use of the simulation model to drive real-world sub-systems: This can be done by Hardware-in-the-Loop (HiL) and Software-in-the-Loop (SiL) simulations. Here the simulations allow to perform functional tests (compliance with protocols) as well as load tests by coupling real-world sub-systems with the virtual sub-systems already existing in the simulation model.

To couple real-world sub-system components with virtual sub-system components two major issues need to be addressed: The abstraction gap needs to be overcome and the time synchronization assured. The abstraction gap results from the typically lower level of detail used in simulation models compared to real-world systems. Models are always build to answer certain questions and therefore they implement only properties necessary to answer this questions. Most often, a sizable amount of structural and behavioral properties can be ignored in that context. However, to couple a real-world system component using HiL/SiL exactly these properties are necessary, e.g. to concisely follow the specification of a given communication protocol.

The second challenge is the synchronization between 'real time' and simulation time. Usually the simulation time should progress significantly faster than reality. This is a desired effect, since the simulation results need to be available as fast as possible—decision support is infeasible when the simulation results do not arrive in time. Nevertheless, it is necessary to synchronize the simulation speed down to real time as soon as data is exchanged with a real-world component for the first time. Up to that point in time the simulation may run at a faster speed to reach a certain system state. Then both time frames need to proceed in step. In addition, it may

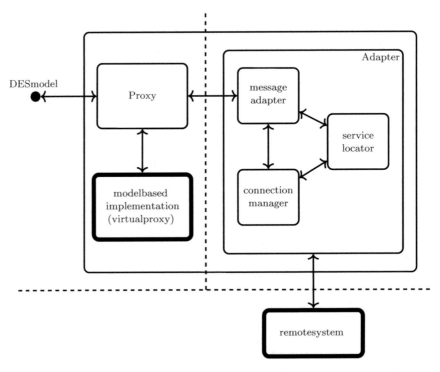

Fig. 2 A proxy pattern [10, 11] is used to implement a sub-system within the model either as a DES model (*left of the dashed line, bold*) or to adapt the discrete event messages to the technical interface used in a real-world remote system (*below the dashed line, bold*)

become necessary to take over the states from the virtual representation of the real system component to the real component (for example open sessions of a server). One technical prerequisite is the possibility to adjust the simulation speed during an ongoing simulation—a (not yet common) feature of the simulation environment.

The typical architecture of a HiL/SiL simulation solution is shown in Fig. 2. The model-based representation of a certain sub-system is permanently or temporarily replaced by a real-world system. A proxy component within the simulation model conveys the requests between the model-based system and the real-world sub-system. In addition, an adapter is necessary to translate the discrete event messages to the technical interface used in the real system. The adapter receives discrete event messages from the simulation model and creates actual network connections (e.g. TCP/IP based). In most cases additional information related to the connection is necessary, since the messages of the simulation model are depicted at a higher abstraction level. In our example, all network connections are managed by the same adapter in parallel and the next section explores the scalability of this approach. In effect this means that a serial-mode simulation environment may become too slow to process requests in a timely fashion, when a large amount of parallel connections are active—even when the simulation speed is normally faster than reality.

3 Performance Issues

To investigate the potential performance issues when coupling a serial simulation model to a parallel real world adapter we created a synthetic benchmark model in MSArchitect [12]. The model mimics the interplay of a simulation model exchanging data with a web-server using thousands of parallel TCP/IP connections simultaneously. The synthetic benchmark model implements a test driver component to open a configurable amount of parallel connections to a real server. The main task of the adapter is to to redirect the data flow from the model to the real server and vice versa, where 1 kB of data is transferred in each communication session with the ability to throttle the bandwidth of each connection e.g. to the bandwidth typical in mobile data networks.

For the analysis we created several setups of synchronous and asynchronous multi-threaded TCP-servers (real-world mock-up server) and TCP-clients (as part of the proxy within the simulation framework MSArchitect) using Winsock [13] and C++ ASIO [14]. The focus was to test if over 10,000 connections can be opened and processed in parallel.

The best results could be achieved by the C++ ASIO based loop-back scenario. Here the client and the server are running on the same computer, more than 10.000 parallel connections (and therefore threads) could be achieved. The whole simulation run was executed within a few seconds using a typical multi-core server. In a second setup the TCP-server sub-system was moved to its own dedicated computer accessible via a LAN connection. This reduced the number of parallel connections to approximately 5.500—adding more connections would lead to connections being closed spontaneously. The reasons are unknown so far, whether it is problem of the network, the operating system or deficiency in the implementation of client and server. In general it could be observed that the time necessary to establish a connection was significantly higher than the time necessary to transfer the data (payload).

4 Example of German Toll System

The performance constraints mentioned above depend on the use case. In our example we take the German automatic toll system as use case—a typical example of a toll system based on global navigational satellite systems [15]. The system uses OBUs deployed in the heavy-goods vehicles (HGVs) to determine the tolls. At present almost one million OBUs are deployed, communicating with the centrals system to transmit toll data or to download updates to the geo and tariff data and the OBU software. In the past we have developed a realistic DES model for the most important processes of the toll system (see [9] and references therein).

For many purposes the simulation model is sufficient even if it retains a certain level of abstraction. In our case the system size is at a scale of 1:1, communication bandwidth and latency are included in the simulation model. However, communication processes are not modeled in detail—neither the processing steps nor the exchange of TCP/IP packets is included in the model. Even with these limitations the simulation model can be applied to predict the dynamic system behavior—either of the existing system under normal operating conditions or beyond the design specification. New processes and systems can be included in the model and put into the 1:1 operational environment to predict e.g. the system sizing or the resource consumption.

For software-in-the-loop simulations the existing model would need to be expanded first to include all process steps and then interface with the relevant communication protocols—most notably communicating via TCP/IP. The interface to TCP/IP based connections is the natural boundary where a real-world system—e.g. the central server responsible for communicating with the OBU fleet—could be interfaced with the simulated behavior of the OBU fleet. Section 2 explained the technical pre-requisite: The ability to translate between an event-driven simulation model and thousands of parallel TCP/IP connections to an existing communication server.

To illustrate the necessary changes to the simulation model we look at the most common communication process used in the toll system (see Fig. 3): The exchange of a status message when no further activities are required. So far the simulation model contains realistic triggers for the process but summarizes the process by including only the beginning and the end of the process (steps A and E in Fig. 3) with the typical time taken for and data transmitted during a connection (a histogram of the overall connection duration is given in Fig. 4).

SiL simulations would need to implement the complete process and to interface with the TCP/IP based communication. Figure 3 indicates the minimum complexity required—even a simple exchange of a status message involves many steps: In the

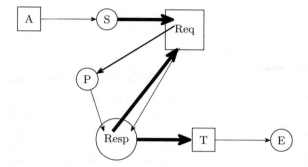

Fig. 3 Process flow for the most common communication process in the toll system example—the status message exchange. Rectangles (*circles*) denote state changes due to client (server) activities, the bold arrows indicate activities with longer duration. The process starts with a TCP/IP connection (A) and the spawning of a server-side communication thread (S), proceeding through the activities request (Req), processing (P) and response (Resp) before the connection is terminated (T) and the thread ends (E)

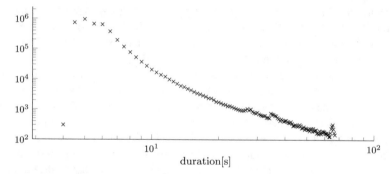

Fig. 4 Histogram of the duration of the status message exchange process (time elapsed between steps A and E in Fig. 3)

example the OBU initiates the communication by establishing a TCP/IP-connection to the central server (step A) which in turn forks a new thread dedicated to the newly established communication channel (step S).

The actual process consists of a simple loop: A request is received (Req) by the server, processed (P) and a response sent to the OBU (Resp). Even with a negligible message size the latency inherent in mobile data networks leads to delays of several seconds (indicated by the bold arrows), wheres server-side processing is typically very fast (on the order of milli-seconds up to a few hundred ms for steps including storage or encryption). To reach this level of detail in the simulation model, access to the technical processes and protocol specifications is needed. Yet—depending on the use case—it might still not be necessary to implement all details within the SiL simulation model, e.g. when the central billing systems are not covered by the model, many details concerning the correct billing data can be omitted from the simulation model.

In practice, access to real-world statistical data is also needed: In fact, in our example the process steps in Fig. 3 are reconstructed from 'process mining' (using the term of [16]) available data for the most common process—a step necessary to build statistical tables of e.g. the time needed to finish one process step. Figure 4 gives an impression of the typical findings when a large fleet of OBUs communicates with the central system: On average a connection is established for a duration of about 6 seconds—stemming mostly from the three message exchanges indicated by bold arrows in Fig. 3, each taking the typical round-trip time observed in GPRS networks [17]. Yet it is obvious that the probability distribution is not a Gaussian distribution: There is a sizable percentage of connections that takes considerably longer than presumed (for transmitting a few hundred bytes over mobile data networks). The log-log plot in Fig. 4 suggests a power law extending to durations of about 60s where a pronounced drop-off occurs (not visible in the figure).

This behavior is typical for TCP/IP connections over high-latency networks [18]: With time, some connections recover from temporary network outages unless a connection time-out (in our case at 60s) closes the connection first.

Regarding a SiL simulation model the empirical distributions need to be incorporated for several reasons: First and foremost the high percentage of connections with considerably longer duration translates into more open connections to be handled by the central server. In reality this behavior is the "real" system requirement—that can only be formulated with prior knowledge and tested by dedicated simulation models as test driver. Furthermore the empirical distribution indicates the scale at which the simulation model needs to respond in "real-time". An example is the message transfer taking at least one second and on average two seconds—the delays introduced by the scheduling of the discrete events must be negligible on this scale.Of course, the system could contain many more processing steps with more stringent timing constraints. Especially server-side computing, e.g. access to persistent storage or delays due to dedicated cryptographic routines could be simulated but will operate possibly orders of magnitude faster.

In reality the proxy pattern used to translate between the discrete event simulation and the TCP/IP connections (see Sect. 2) serializes the communication over thousands of parallel connections into a single future events list to be processed sequentially by the DES kernel. Therefore the "real time" requirement is equivalent to processing all simultaneous events in the future events list within the allotted time frame. At least for this model element we expect that a parallelization is necessary.

5 Summary

Executable requirements are the underlying idea of the simulation driven design approach: In an extension to the test driven development approach, SDD puts every development effort into the context of the whole system running under normal operating conditions. Of course, the approach incurs additional expenses—a simulation model needs to be developed and maintained—yet it aims to shift many tests to the

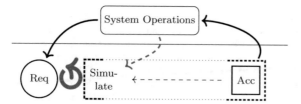

Fig. 5 Simulation driven development shifts the acceptance testing (acc)—at least partially—to the requirements definition phase (req). At all times the simulations put the system under construction into the operational context

very beginning of the software development process. In many respects the operational performance of the simulation model and the tool set determines the applicability of SDD: Requirements from a repository need to be connected to simulation models and model elements, e.g. to keep track of simulation results and for tracing the coverage of the simulation model. Furthermore simulations of large-scale systems at a scale of 1:1 require a high simulation performance (Fig. 5).

Taking a common use case—the simultaneous communication of many slow client computers with a central server—we have shown that a discrete event simulation model is capable to drive a real-world server with many thousand connections in parallel. The remaining challenge was illustrated using the German automatic toll system: In addition to serving a sufficient number of parallel connections, a realistic simulation must not introduce noticeable delays. As a consequence it becomes necessary to parallelize the DES model to keep the 'artificial' delays originating from the model execution at a negligible level.

References

1. K. Petersen, M. Khurum, and L. Angelis, "Reasons for bottlenecks in very large-scale system of systems development," *Information and Software Technology*, vol. 56, no. 10, pp. 1403–1420, Oct. 2014. doi: 10.1016/j.infsof.2014.05.004
2. E. Woods, "Operational: The forgotten architectural view," *IEEE Software*, vol. 33, no. 3, pp. 20–23, May 2016. doi: 10.1109/MS.2016.86
3. B. Pfitzinger and T. Jestädt, *IT-Betrieb: Management und Betrieb der IT in Unternehmen.* Springer, 2016. ISBN 978-3-642-45193-5. doi: 10.1007/978-3-642-45193-5
4. K. Beck, *Test-driven development: by example.* Addison-Wesley Professional, 2003.
5. "Manifesto for agile software development," [accessed 29-Jul-2015]. [Online]. Available: http://www.agilemanifesto.org/
6. J. Roche, "Adopting DevOps practices in quality assurance," *Communications of the ACM*, vol. 56, no. 11, pp. 38–43, 11 2013. doi: 10.1145/2524713.2524721
7. S. Newman, *Building Microservices.* O'Reilly Media, Inc., 2015. ISBN 978-1-4919-5035-7
8. B. Pfitzinger, T. Baumann, D. Macos, and T. Jestädt, "Using parameter optimization to calibrate a model of user interaction," in *Proceedings of the 2014 Federated Conference on Computer Science and Information Systems*, ser. Annals of Computer Science and Information Systems, M. P. M. Ganzha, L. Maciaszek, Ed., vol. 2. IEEE, 9 2014. doi: 10.15439/2014F123. ISBN 978-83-60810-58-3. ISSN 2300-5963 pp. 1111–1116.

9. T. Baumann, B. Pfitzinger, D. Macos, and T. Jestädt, "Enhanced simulation performance through parallelization using a synthetic and a real-world simulation model," in *Proceedings of the 2015 Federated Conference on Computer Science and Information Systems*, ser. Annals of Computer Science and Information Systems, M. Ganzha, L. Maciaszek, and M. Paprzycki, Eds., vol. 5. IEEE, 9 2015. doi: 10.15439/2015F226 pp. 1335–1341.

10. E. Gamma, R. Helm, R. Johnson, and J. Vlissides, *Design Patterns: Elements of Reusable Object-oriented Software*. Addison-Wesley Longman Publishing Co., Inc., 1995. ISBN 0-201-63361-2

11. D. Alur, D. Malks, J. Crupi, G. Booch, and M. Fowler, *Core J2EE Patterns (Core Design Series): Best Practices and Design Strategies*, 2nd ed. Sun Microsystems, Inc., 2003. ISBN 0131422464

12. Andato GmbH & Co. KG, "MSArchitect," [accessed 10-Dec-2012]. [Online]. Available: http://www.andato.com/

13. B. Quinn, *Windows Sockets Network Programming*, 2nd ed. Addison-Wesley Longman Publishing, 1998. ISBN 0201183943

14. C. Kohlhoff, "Boost.Asio," 2015, [accessed 02-May-16]. [Online]. Available: http://www.boost.org/doc/libs/1_60_0/doc/html/boost_asio.html

15. A. T. W. Pickford and P. T. Blythe, *Road user charging and electronic toll collection*. Artech House, 2006. ISBN 978-1-58053-858-9

16. W. Aalst, *Process Mining: Data Science in Action*. Springer, 2016. ISBN 978-3-662-49851-4. doi: 10.1007/978-3-662-49851-4

17. G. Xylomenos, G. C. Polyzos, P. Mahonen, and M. Saaranen, "TCP performance issues over wireless links," *IEEE Communications Magazine*, vol. 39, no. 4, pp. 52–58, Apr 2001. doi: 10.1109/35.917504

18. M. C. Chan and R. Ramjee, "TCP/IP performance over 3G wireless links with rate and delay variation," *Wireless Networks*, vol. 11, no. 1-2, pp. 81–97, Jan. 2005. doi: 10.1007/s11276-004-4748-7

Cognitum Ontorion: Knowledge Representation and Reasoning System

Paweł Kaplanski and Pawel Weichbroth

Abstract At any point of human activity, knowledge and expertise are a key factors in understanding and solving any given problem. In present days, computer systems have the ability to support their users in an efficient and reliable way in gathering and processing knowledge. In this chapter we show how to use Cognitum Ontorion system in this areas. In first section, we identify emerging issues focused on how to represent and inference knowledge. Next, we briefly discuss models and methodology of agent-oriented system analysis and design. In the third section, the semantic knowledge management framework of the system is reviewed. Finally, we recapitulate by discussing the usability of Ontorion based on a case study, in which an instance of software process simulation modelling environment is executed and further discussed. In the last section, we provide future work directions and put forward final conclusions.

1 Introduction

"Knowledge is power"—this phrase is often attributed to Francis Bacon. Indeed, a lot of human effort and material resources have been exploited in preserving knowledge. Indubitably, humans have been always using many different forms to express and share knowledge with others (e.g. drawings, symbols, words and numbers, which can be easily encoded in computer memory). To understand essence of our approach, we shall now focus on some formal methods of knowledge representation, which are the prominent research domain of artificial intelligence (AI).

P. Kaplanski (✉) · P. Weichbroth
Faculty of Management and Economics, Department of Applied Informatics in Management, Gdansk University of Technology, Gabriela Narutowicza 11/12, 80-233 Gdansk, Poland
e-mail: pawel.kaplanski@zie.pg.gda.pl

P. Weichbroth
e-mail: pawel.weichbroth@zie.pg.gda.pl

© Springer International Publishing AG 2017
T. Pełech-Pilichowski et al. (eds.), *Advances in Business ICT: New Ideas from Ongoing Research*, Studies in Computational Intelligence 658,
DOI 10.1007/978-3-319-47208-9_3

In the AI canon, knowledge seems to be always defined in a strictly functional way. However, from all incoming questions to the mind of an sedulous reader, which one would be the first to reveal the question: what is knowledge? or "when data or information become knowledge?"—This might be the difficult question to answer, if we take into account diffusion of these three artefacts. Bearing in mind the fact that a computer "brain"—central processing unit (CPU) is principally able to process "only" sequences of bits, where a single bit is represented by two exclusive numbers: 0 (zero) or 1 (one). In consequence, at the moment as far as we know, human knowledge is represented "only" by facts and rules in computer memory. A distinct fact (or a set of facts), represented by a sentence (or a set of sentences), is used in deductive reasoning. A single rule (or a set of rules), expressed in a form: *if* → *then*, may be a logic or be inductive in its genesis. Elements of particular knowledge (intra- or interconnected facts or rules) are often named as *"knowledge chunks"* [1]. A single chunk is commonly attached to an exclusive agent (an independent and separate application unit).

In the beginning, AI research investigated how a single agent can exhibit singular and internal intelligence. However, in recent years, we can observe higher interest in concurrency and distribution in AI, which have been reflected in its name: distribution artificial intelligence (DAI). In the literature, it has been grounded that DAI can be considered in two primary research areas: distributed problem solving (DPS) and multi-agent (MA) systems.

Some successful application of agent-oriented architecture can be pointed in decision support systems (DSS) in the area of the discovery of stock market gamblers patterns [2, 3], web usage mining [4, 5] and ecommerce environments [6]. As far as we know, it is not a straightforward task to coordinate knowledge, goals and actions among a collection of autonomous agents. Thus, to perform such complex tasks, we shall use an application which provide such language to codify knowledge in the form of ontology, which can be easily understood by a user. Our choice is the Cognitum Ontorion, developed by software vendor from Poland. The Ontorion Fluent Editor is a comprehensive tool for editing and manipulating complex ontologies, developed in Controlled Natural Language. The Controlled English is a main built-in feature which makes this tool more suitable to others XML-based OWL editors. It is also supported via Predictive Editor that prohibits one from entering any sentence that is grammatically or morphologically incorrect and actively servers a help to the user during sentence writing. For each edited OWL file, a taxonomy tree is built upon data from this file and all included ontologies. Each tree is visualized, using user-friendly palette of colors and divided into four parts: thing (shows "is-a" relations between concepts and instances), nothing (shows concepts that cannot have instances), relation (shows relations hierarchy between concepts and/or instances) and attribute (shows attributes hierarchy).

2 Reasoning in Agent-Oriented Design and Analysis. A Hybrid Approach

There is a long history of symbolic reasoning usage in order to provide intelligent behaviour in Multi-Agent and Simulation systems (MASS). Deductive Reasoning Agents, which use logic to encode a theory defining the best action to perform in a given situation, are the "purest" in terms of their formal specification. Unfortunately, they suffer from all the practical limitations of formal representation: firstly, the complexity of theorem proofs (it may even lead to undecidable statements) and secondly, the boundaries of expressivity formed by core knowledge representation attributes (e.g. monotonicity of knowledge, open world assumption).

Making deductive reasoning requires the selection of underlying logics that support the nature of agents. It is worth mentioning that the most prominent implementations of deductive reasoning agents are based on intentional logics like formal models of intention logics [7] (e.g. Belief—Desire—Intention, BDI), which take into account some subset of the Saul Kripke modal logic [8].

Problems with symbolic reasoning led to the establishment of the "reactive agent movement" in 1985, revealing an era of reactive agent architecture. The reactive agent movement manifested in the form of requirements for so-called behavior languages [9]:

1. Intelligent behavior can be generated without explicit representations of the kind that symbolic AI proposes.
2. Intelligent behavior can be generated without explicit abstract reasoning of the kind that symbolic AI proposes.
3. Intelligence is an emergent property of certain complex systems.

Reactive agents are nowadays well recognized but still they lack formal foundations and therefore these kind of MASS are very hard to analyze with formal methods and tools. Nevertheless, the reactive agent movement resulted in Agent Oriented Programming (AOP), e.g. JADE [10], which is currently considered as a step beyond Object Oriented Programming (OOP) in Software Engineering.

A novel approach to designing MA systems—Hybrid Agent Architecture, attempts to combine the best of symbolic and reactive architectures. The system itself is built up of at least two subsystems: (1) a symbolic world model that allows plans to be developed and decisions made and (2) a reactive engine which is capable of reacting to events without involving complex reasoning.

As an example let us consider Ferguson's "TouringMachine" [11], which fits into the definition given above. Ferguson defines: *"The TouringMachine agent architecture comprises three separate control layers: a reactive layer, a planning layer, and a modelling layer. The three layers are concurrently-operating, independently-motivated, and activity producing: not only is each one independently connected to the agents sensory apparatus and has its own internal computational mechanisms for processing appropriate aspects of the received perceptual information, but they are also individually connected to the agents effectory apparatus to which they send,*

when required, appropriate motor-control and communicative action commands".
Whereas, general framework of MA system requires definition of three layers: (1)
agent layer, that describes inputs and outputs, internal structure, goals and tasks, (2)
governing layer, that contains rules (priorities, authorities) and defines relationships
between agents (inheritance and roles) and (3) cooperation layer, that embodies the
interactions between agents.

We present a novel approach to hybrid agent architecture, which is implemented
on top of a scalable Knowledge Representation and Reasoning (KRR) system. KRR
allows each environment to be described formally as well as giving the possibility to
build a reactive agent system based on a knowledge base (KB) triggering subsystem.
Moreover, it provides agents with synthesis tools.

Here, we consider a reactive agent that is able to maintain its state [12]. This
agent has an internal stand-alone data structure, which is typically used to record
information about the state and history of the environment and to store a set of all
the internal states of an agent (Fig. 1).

The perception of an agent is realized in its see function if the function is time-
independent. The agents action selection is defined as a mapping from its internal
states to actions. The next function maps an internal state and percept to an internal
state. The abstract agent control loop is then:

1. Start with the initial internal state $s \leftarrow s_0$.
2. Observe the environment state e, and generate a percept $p \leftarrow see(e)$.
3. Update the internal state via the next function $s \leftarrow next(s, p)$.
4. Select an action via the action function $a \leftarrow action(s)$.
5. GOTO 2.

The environment state e is (in hybrid architecture) given by the symbolic system here we use ontology to encode it. The function next is one that needs to be
implemented either by the programmer or by an automated process. In the first case,
however, it is hard to distinguish such an agent from a (considerably complex) object
oriented program (formerly, active-object design pattern implementation [13]). On
the other hand, automated agent synthesis is an automatic programming task: given
an environment, lets try to automatically generate an agent that succeeds there. The

Fig. 1 An agent with its
internal state

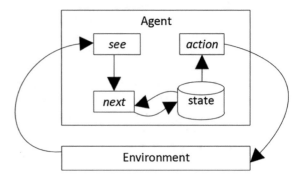

synthesis algorithm should be both sound and complete. Sound means here that the agent will succeed in the given environment once it is correctly constructed, and completeness guarantees the possibility to create the agent for the given environment.

The hybrid approach allows us to build a semi-formal foundation for MASS that allows for a sound and complete synthesis of agents as long as their definition fits into the expressivity frame of underlying logic and if the underlying logic has the reasoning task to be sound and complete itself. This is true for Description Logic (DL) [14] the foundation for OWL [15], therefore we selected OWL compliant KRR.

3   Ontorion Architecture

Modern Scalable Knowledge Management Systems give the possibility to use KRR in a similar way as we tend to use RDBMS. We have focused on a KRR system the functionality of which allows a user-interface in natural language to be implemented and used. Ontorion [16] is a Distributed Knowledge Management System that allows semi-natural language to be used to specify and query the knowledge base. It also has a built-in engine trigger which fires the rules each time if the corresponding knowledge is modified. Ontorion supports the major W3C Semantic Web standards: OWL2, SWRL, RDF, SPARQL. Ontologies can easily be imported from various formats, exported to various formats, and accessed with SPARQL [17]. Solutions built on top of Ontorion can be hosted both in the Cloud and On-Premise environments.

By design, Ontorion allows one to build large, scalable solutions for Semantic Web. The scalability is realized by both the noSQL, symmetric database Cassandra [18] and the internal ontology modularization algorithm [19]. Ontorion is a cluster of symmetric Nodes, able to perform reasoning on large ontologies. Every single system node is able to do the same operations simultaneously on data sets it tries to get the minimal suitable ontology module (part) and perform any requested task on it.

The symmetry of the architecture of the cluster provides system scalability and flexibility Ontorion can be deployed and executed in a computing cloud environment, where the total number of nodes can be changed on request, depending on user requirements.

The fundamental algorithm in a KRR system such as Ontorion ought to reason over description logic selected as a foundation for OWL called SROIQ(D) [20], and should be able to process complete or selected segments of ontologies.

If performance is more important than expressivity power, then it is possible to switch Ontorion into OWL-RL+ mode. OWL-RL+ mode is constructed in a similar way to how it was first implemented in DLEJena [21]. The reasoning process in OWL-RL+ mode remains in SROIQ(D) for the T-Box, while for the A-Box the reasoning is based on the OWL-RL ruleset.

Furthermore, the modular separation of complex ontologies also allows the reasoning process to be partitioned, which can be performed on knowledge modules

Fig. 2 The Ontorion KRR

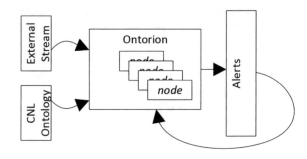

Fig. 3 General form of the
move function

```
If  ...  then  ...  execute  <?
              Move(agent,"current-state", message,
                      "expected-message",()=>
          {
                      /* the agent action */
                      return "new-state";
          });
    ?>.
```

(independent pieces of knowledge) in parallel, at the same time on separate machines. In other words, the modularization algorithm is scalable and traceable.

In Ontorion, conclusions that are the results of new incoming knowledge can fire triggers at extension/reactive points (Fig. 2). On the other hand, if some chunk of knowledge meets a set of predefined conditions, a knowledge modification trigger executes the procedures responsible for interaction with external systems (e.g. sending a notification using an SMTP server). We observed that knowledge modification triggers allow the Hybrid Agent MASS to be built on top of the Ontorion KRR (Fig. 3).

The underlying storage for Ontorion is the BigTable [22] implementation (namely Cassandra), which is able to maintain a petabyte of data. Together with an analytic cluster e.g. Hadoop [23] it forms a BigData solution. In these terms, we can consider Ontorion as a BigKnowledge solution and our Agent application as a BigAgent, which together constitutes a highly scalable hybrid agent infrastructure.

Distributed systems lack the one common model of time. In a distributed environment, time is relative and the serialization of events requires an internode negotiation algorithm. In modern distributed noSQL databases like Cassandra 2.0, the serial time model can be preserved with the use of the Paxos algorithm [24], which allows a distributed atomic "Compare and Set" (CAS) functionally to be efficiently implemented.

CAS in Ontorion is used to maintain the Agent state in a coherent way, therefore the existence of CAS is critical for proper system functioning. CAS also provides a way to make the Agent System fault-tolerant by transactional-queue implementation, which is crucial for long-running simulations.

4 Programming Agents with Natural Language

An important, novel feature of Ontorion, among other KRR systems, is the ability to describe knowledge and interact with the user in semi-natural language. The language represents the family of controlled natural languages that is expressive enough to describe OWL. Controlled natural language (CNL) is a subset of natural language with a reduced grammar and vocabulary, which in this case translates directly to logic with formal semantics capabilities.

In general, controlled natural language should be unambiguous and intuitive, ultimately forming an easy way for human-machine interaction (understandable by humans, executable by machines). Due to its limitations, it needs to be supported by a predictive (structural) editor which is Ontorion FluentEditor tool [25]. The other, well-known implementation of a controlled natural language is "Attempto Controlled English (ACE)" [26], developed by the University of Zurich. However, the origins of CNL can be found in the famous novel by George Orwell: "1984", where he discusses the NEWSPEAK a controlled language. The most used industrial implementations nowadays are Domain Specific Language (DSL) (implemented as a part of the Drools project) [27] and Semantics of Business Vocabulary and Rules (SBVR) [28], whereas CNL allows the representation of BPML diagrams.

In Ontorion Fluent Editor controlled language is equipped with formal semantics expressed in logic. General groups of sentences are allowed which include:

1. **Concept subsumption**, represents all cases where there is a need to specify (or constrain) the fact about a specific concept or instance (or expressions that evaluate the concept or instance) in the form of subsumption (e.g.: *Every dog is a mammal, Paul has two legs or One dog that is a brown-one has red eyes*).
2. **Role** (possibly complex) inclusion specifies the properties and relationships between roles in terms of the expressiveness of $SROIQ^{(D)}$ (e.g.: *If Jane loves something that covers Y then Jane loves-cover-of Y*).
3. **Complex rules**; If [body] then [head] expressions that are restricted to the DL-Safe SWRL subset [29] of rules (e.g.: *If a scrum-master is-mapped-to a provider and the scrum-master has-streamlining assessment-processes-sprints-level equal-to 2 then the provider has-service-delivery-level equal-to 1 and the provider has-support-services-level equal-to 2*).
4. **Complex OWL expressions**; the grammar allows the use of parentheses that can be nested if needed in the form of (that) e.g.: *Every mammal is something (a male or a female or a hermaphrodite)*.
5. Above knowledge modification triggers that have the form of: *If* P *then for-each* P *execute* Q, where P is a premise and Q a consequence. Premise P is an expression that evaluates a set of connected instances that fulfill some conditions, while the consequence Q is a procedure written in C# programming language (e.g.: Fig. 5).

5 The Definition of the Multi-agent System in Terms of Knowledge Management System Triggers

Here, we present a modern scalable KRR as a foundation for MASS. The discussed KRR (Ontorion) enables the specification of knowledge-modification triggers in the form of reactive rules: *if* → *action*. Ontorion knowledge modification triggers allow the knowledge itself to be modified and therefore it is possible to build a set of triggers here that are fired continuously. A reactive trigger like this breaks the decidability of the underlying knowledge base and, as a consequence, KRR tasks based on Ontorion are decidable only if all deductive rules are DL-Safe (e.g. they are SWRL equivalent) otherwise these tasks are non-decidable.

The above property of system modification triggers is very useful for the development of the hybrid MASS. The hybrid agent paradigm can be adapted, by using triggers, even if the environment is modelled in Ontorion as an OWL Ontology, with all its limitations (e.g.: lack of modality or time representation). We can define agents here as OWL individuals. The behaviour of the agents is implemented in reactive rules called moves. These rules combine the see, next and action functions discussed earlier (the reactive, state based and abstract model of an agent). Moreover, agent-individuals and "ordinary" OWL individuals are different as agent-individuals are equipped with a transactional, CAS protected internal state, represented by related data-values.

A single move function as a parameter takes a percept which is a result of a reasoning process (here, we consider reasoning as an implementation of the abstract see function) over the current state of the environment. In the implementation of this function a CAS operation is used to preserve transactional semantics. The move function is only activated if the perceived-message is equal to the expected-message.

There is not one single agent that activates on perceived-message—it can be any agent that fulfills the rule premise, therefore a rule conclusion can be reused by many agents and the overall execution result of the system is non-deterministic. Messages are transferred between agents using a distributed message queuing system, managed by the KRR.

A single agent is determined by its state and all the move functions that can be ever executed in the context of its state, therefore here, the agent synthesis process, is a process of the assignment of the move functions to the single agent-individual.

The reactive-rule bodies of the move function determine the specific environment state that allows the system to assign the function to the agent; however, the overall behaviour of MASS is non-deterministic. This is due to the fact that the concrete run (the MASS run) needs the selection of agent instances made in runtime and runtime (in opposition to reasoning time) is a part of the reactive model that is nondeterministic by nature. The non-deterministic selection of choices, often by use of pseudo-random number generators, and the parallel execution of different threads, are required by underlying technologies to provide an efficient computational model.

Therefore, we need to keep in mind, that simulations based on the reactive/hybrid approach are non-deterministic. In this case, we have to perform a large set of experiments with the same initial state and then use analytical tools and methods to verify the statistical hypothesis.

6   The Scalability of KRR Oriented MASS

In practical scenarios, when it comes to simulating large societies of objects, it is necessary to simulate a large amount of agents at the same time. From the technological perspective, currently, we can model large societies of people. Nowadays, the existence of $7\ddot{O}109$ beings can be encoded in less than 1 GB of memory. If we encode a single human being as a 1 kB vector of bits then we can store an entire population in a single modern hard drive at a relatively low cost. Working with cloud-based environments, we can hire thousands of computers for a few hours with a similar amount of money. Therefore MASS scalability the ability of the system to scale together with the size of the problem—is regarded as a mandatory and critical property.

Ontorion is a scalable KRR system, approximately associated with the size of maintained knowledge due to modularization algorithms embedded, whereas Cassandra, as the underlying storage solution, is scalable by its design.

In distributed systems, task synchronization is a burden and sometimes even an obstacle. Task distribution over a set of physical machines demands synchronization protocols. Satisfyingly, the Cassandra database has the Paxos protocol implemented, which allows a global CAS functionality to be implemented. The ability of agents to modify particular chunks of knowledge indicates influence on the surrounding environment as well. What can be seen as a common task in terms of RDBMS (e.g. some database modification), might have large and complex implications in terms of a distributed knowledge base. Given the subset of First Order Logic (FOL), we deal with the monotonic knowledge model. The monotonicity implies that there is no impact on the overall meaning when the order of adding knowledge is one way or another.

When we modify knowledge the problem is somehow more complicated besides agents tend to modify knowledge very often. The cost of knowledge modification depends on its level, scale and size. The relation between knowledge generality and its modification cost is positive as a result of the replacement of all revalued conclusions. Moreover, knowledge modification triggers, used to implement the next functions, break the open world assumption (OWA) [14]. This effect is caused by their ability to modify knowledge depending on the "known" parts of the knowledge. An agent may learn knowledge even when it stays in contradiction to what it already knows. In addition, knowledge modification triggers break the monotonicity of the knowledge base. Therefore, the order of agent Next firing is significant in terms of the final knowledge base shape. As previously mentioned, simulations based on the reactive/hybrid approach are non-deterministic due to both the distributed system properties and the internal non-determinism of the reactive agent system.

7   Experimental Setup

Software Process Simulation Modelling (SPSM) [30] is widely used nowadays to
support planning and control during software development. MA systems play a very
important role here as they naturally can be used to simulate social behaviors in the
software testing phase. In our approach, the SPSM is divided into two components:
ontology and knowledge modification triggers. In the example given below (see
Fig. 4), the ontology defines (with CNL) the core concepts such as: competency,
task, developer, manager.

   We also defined agent-rules by making use of knowledge modification triggers
(see Figs. 5 and 6). Those triggers implement the following scenario: a developer
with certain competencies starts to realize a task. After the task is finished, new
knowledge about the task realization process is added, and a "Busy" state is set on
the developer. The second trigger is fired when the task is finished and a "Ready"
state is set back.

   Every time the environment contains a situation where one task is dependent on
the other, finished task, we execute the trigger that forces previous triggers to start a
simulation.

   Here, we use the once function, which ensures that the execution of the trigger
happens exactly once (note that this is not a trivial task, not executed in a stand-alone
system but in a distributed system which requires dispersed CAS operation). The last
step is to define the simulation entry point (see Fig. 7).

```
Cpp–Programming is a competency.
Java–Programming is a competency.
. . .
Task–0 is a task.
Task–1 is a task.
Task–1 is–dependent–on Task–0.
Task–1 requires–competency Cpp–Programming.
Task–1 has–estimated–realization–md equal–to 500.
Task–2 is a task.
Task–2 is–dependent–on Task–1.
Task–2 requires–competency Java–Programming.
Task–2 has–estimated–realization–md equal–to 500.
. . .
Anna is a developer.
Anna has–competency Cpp–Programming.
Anna has–competency Java–Programming.
John is a developer.
John has–competency Java–Programming.
John has–competency Web–Programming.
. . .
```

Fig. 4 Configuration of the SPSM environment

```
If a task−realization−query requires−competency a
competency and a developer has−competency
the competency and the task−realization−query has−origin
a task then for the task−realization−query and the
developer and the task execute <?
        Move( developer , "Ready",
                task_realization_query , "Programming", ()=>
            {
// read the realization time
var realizationTime =
        ( from v in Values where
        v . source= =InstanceDL ( task ) &&
        v . datarole= ="have−estimated−realization −md"
        select v . value ) .
        FirstOrDefault () .
        SetConsistencyLevel ( ConsistencyLevel . Quorum ) .
        Execute () ;
// create the wake−up message
                var msgid = CreateMessage ( developer ,"WakeUp" ,task ) ;
// delayed (by the realization time) modification //
of KB
        KnowledgeInsertWithDelay (
                msgid + " is a wake−up−message . "+
                msgid + " has−origin " + developer + " . " +
                msgid + " has−task−realization −query "
                        + task_realization_query + " . " +
                msgid + " has−target " + developer + " . " ,
        int . Parse ( realizationTime ) ) ;
// mark the agent state as busy
                return "Busy";
        }) ;
?>.
```

Fig. 5 The *move* function written in C# is fired when the developer is ready and fit for the given task

```
If a wake−up−message has−target a developer and the
wake−up−message has−origin the developer and the wake−up−message
has−task−realization −query a task−realization −query and the
    task−realization−query has−origin
a task then for the wake−up−message and the developer and the
    task−realization −query
and the task execute <?
        Move( developer ,"Busy",
                wake_up_message , "WakeUp" ,()=>
            {
// modify the status of task
                KnowledgeInsert ( task+" has−status Done . ") ;
// mark the agent state as ready
                return "Ready";
        }) ;
?>.
```

Fig. 6 The *move* function written in C# is fired when the developer is done with the task

```
If  a  start−event  exists  then  for  the  start−event  execute
<?
        Once("Lets  the  simulation  start .",  ()=>
        {
                CreateAgent("Mark","Ready");
                CreateAgent("John","Ready");
                CreateAgent("Tom","Ready");
                CreateAgent("Gabi","Ready");
                CreateAgent("Anna","Ready");
                KnowledgeInsert("Task−0  has−status  Done.");
        });
?>.
```

Fig. 7 The simulation entry point

Start−Event is a start−event.

Fig. 8 The simulation entry point

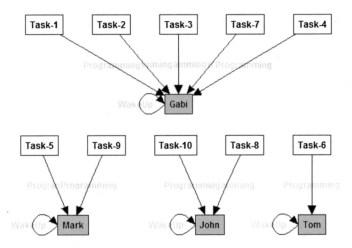

Fig. 9 The result of a particular SPSM simulation (assignment of programmers to tasks)

The start event (see Fig. 8) sets up the agents and defines the initial task for the "Done" state to activate the overall simulation (Fig. 9).

In the above simulation, to make a long story short, in the beginning, we defined three distinct sets: task, competence and individual. Each agent represented a particular individual (developer). Each task required a precise competency and was time specified. This simplified description of the modelled micro-world was given as an input to the system. Next, on a user request the simulation was executed and a set of rules started to be processed due to accomplish a given set of tasks (Fig. 10). Ontorion usability has been evidenced in one of many possible applications. In highly complex systems or projects, it seems that it is a considerable issue to design, estimate and finally test all possible dependency relationships between processes and

```
0,000   Anna   Ready   Programming   Busy    Task-1
1,782   Anna   Busy    WakeUp  Ready  Task-1
3,058   Anna   Ready   Programming   Busy    Task-2
4,380   Anna   Busy    WakeUp  Ready  Task-2
10,239  Anna   Ready   Programming   Busy    Task-11
10,332  Gabi   Ready   Programming   Busy    Task-12
10,598  John   Ready   Programming   Busy    Task-7
10,848  Mark   Ready   Programming   Busy    Task-5
12,431  Tom    Ready   Programming   Busy    Task-6
16,593  Mark   Busy    WakeUp  Ready  Task-5
16,749  Mark   Ready   Programming   Busy    Task-4
19,306  Gabi   Busy    WakeUp  Ready  Task-12
20,254  Gabi   Ready   Programming   Busy    Task-9
26,214  Mark   Busy    WakeUp  Ready  Task-4
30,358  Mark   Ready   Programming   Busy    Task-10
31,228  Gabi   Busy    WakeUp  Ready  Task-9
31,579  Tom    Busy    WakeUp  Ready  Task-6
34,636  Gabi   Ready   Programming   Busy    Task-3
36,639  Mark   Busy    WakeUp  Ready  Task-10
39,322  Mark   Ready   Programming   Busy    Task-8
41,086  Anna   Busy    WakeUp  Ready  Task-11
49,549  Gabi   Busy    WakeUp  Ready  Task-3
49,972  Mark   Busy    WakeUp  Ready  Task-8
58,962  John   Busy    WakeUp  Ready  Task-7
END
```

Fig. 10 The systems console with detailed information of task status, performer and dependencies

their execution sequence. We showed how to optimize the selection of a developers competency to a particular tasks. In this instance, we were able to identify "hidden" bottlenecks and constraints.

The experiments performed on the Ontorion cluster show flexible system scalability, persistent intra-communication duration between nodes and overall system stability. As an illustration, let us present this factual operation. A new node added to the server farm was properly initialized and broadcasted to other nodes and a scheduled job was again distributed. Still, a systematic empirical measurement needs to be made to monitor system behavior, especially when some changes have taken place. A cloud-based environment is our choice due to the obvious benefits of machine virtualization. For Ontorion cluster setup, we used cluster of 3 standard VM nodes. On top of this cluster we executed MASS made of 5 agents.

8 Examples of Ontorion Applications

Due to the limitations of this paper, we only briefly mark a few Ontorion applications which we think give an objective perspective on its functionality. First of all, our system has been successfully deployed as an intelligent semantic tool in a company from the energy sector located in the USA. There are several benefits worth mentioning of customizing Ontorion to this client. Primarily, we were able to semantically describe data sources due to automating the process of infrastructure management. Another interesting application, which has taken place recently in a company in the Aeronautics and Space industry, is a case-based reasoning solution. Here, we combine text mining with a dedicated ontology to mine and structure information residing inside messages, incoming from users, which expose some third-party system errors and defects. Next, to those extracted chunks of information, the inference engine adds relevant tags based on a semantic analysis and together, such enhanced information (someone may even define such rich data as knowledge) is exported to a database (knowledge base). Later, we execute triggers to combine this knowledge with rules and reason a possible set of actions. Nevertheless, an expert is responsible for selecting the final action to be taken due to the solution of the reported problem. In medicine, for the Maria Skodowska-Curie Institute of Oncology we built a highly complex ontology which represents a set of rules describing cancer procedure treatment. First and foremost, we are able to centrally manage all the rules, where, on a users rule change request, with little effort, we just need to modify the semantic description preserving other rules from unnecessary modifications. The most beneficial is the usage of (semi) natural language that is readable for medical oncology experts. Thus in this way they are able to verify the knowledge input.

In the production sector, we have developed a common dictionary for different actors to communicate and collaborate world-wide. Actors include employees (located in different countries and continents) and heterogeneous information systems (from different software vendors). The deployment of semi-natural language, integrated inside the Fluent Editor (Ontorion ontology editor), proved to be a tool, on the one hand, easy to understand and use by its users, and on the other hand, easy to configure and maintain for administrators during systems (applications) integration.

Finally, we can use Ontorion to manage authorization and authentication, versioning, auditing and what distinguishes us from the competition is collaborative ontology engineering. Moreover, we are able to deploy a solution for semantic enhance searching which, based on taxonomy, efficiently improves sharing and searching for information in response to a user query, and interpreting strings of words not only using statistical techniques, but in the sense of logical connections existing between them. On the other hand, such a taxonomy can be seen as an asset of organization knowledge, which can be used to acquire knowledge from individuals and next to preserve it in a formal way.

9   Future Work and Conclusions

First, we plan to explore the scalability of MASS created in the way described in this paper. Even if Ontorion itself is a scalable solution, it is not obvious that the knowledge and inter-agent communication via a distributed queue will have the property of scalability too. Secondly, we want to explore an automated agent synthesis, based on theorem proofers. The synthesis should select rules in an adaptive way from the set of available rules by activating them with some threshold. In this paper, we showed that the Ontorion server is able to execute, maintain and control massive simulations based on a hybrid MASS approach. The resulting MASS fits well into the definition of a Hybrid MASS. The perception of an agent is realized by description logic theorem proofers (reasoners). Agents are modelled as instances equipped with a relevant time model, which allows them to interact with the environment and with each other. Inter-agent communication is realized by messages that together with the environment are represented by an ontology managed by the server. Finally, actions performed by agents are encoded in a particular programming language, which brings our approach close to AOP.

An agent synthesis benefits from a formal reasoning engine (being a central component of KRR) and is based on an action-selection procedure. An expressive and distinct Ontorion functionality is the ability to encode agent logics in semi-natural language in the interest of a less professional user. We also observed that this allows end users to understand actions taken by an agent, even if a user is not well trained in formal representation systems.

Obviously, we are aware of some obstacles which can be pointed out in the presented simulation. The key issue is to verify the usability and then estimate the degree of functionality adoption on the client side. We have also not used any technique of knowledge verification and validation [31]. On the other hand, we presented some Ontorion applications in the medicine, aerospace and production industries which were positively evaluated and are still in use. Based upon preliminary feedback from our clients, we think that the presented system, as a multi-agent simulation platform, is a promising prospect not limited to any particular industry or purpose.

References

1. J. A. Jakubczyc and M. L. Owoc, "Contextual knowledge granularity," in *Proceedings of Informing Science & IT Education Conference (InSITE)*, 2011, pp. 259–268.
2. M. Bac, J. Korczak, A. Fafula, and K. Drelczuk, "A-trader - consulting agent platform for stock exchange gamblers." in FedCSIS, 2012, pp. 963–968.
3. J. Korczak, M. Hernes, and M. Bac, "Risk avoiding strategy in multiagent trading system." in FedCSIS, 2013, pp. 1119–1126.
4. Weichbroth P.: The system framework for profiling the content of web portals. W: Korczak J. (red.): Business Informatics. Data mining and Business Intelligence. Research Papers of Wroclaw University of Economics. Wydawnictwo Uniwersytetu Ekonomicznego we Wrocawiu, Wrocaw 2009, pp. 186–193.

5. P. Weichbroth and M. Owoc, "A framework for web usage mining based on multi-agent and expert system an application to web server log files," Prace Naukowe Uniwersytetu Ekonomicznego we Wrocawiu, no. 206, pp. 139–151, 2011.

6. M. Owoc and L. Piasny, "Adaptacyjne systemy agentowe we wspólczesnym środowisku e-commerce", Problemy Zarzadzania, 13, 2015.

7. A. S. Rao and M. P. George, "BDI agents: From theory to practice," in Proceedings of the First International Conference on Multi- Agent Systems (ICMAS-95), 1995, pp. 312–319. [Online]. Available: http://www.agent.ai/doc/upload/200302/rao95.pdf.

8. S. Kripke, Naming and necessity. Harvard University Press, 1980. ISBN 9780674598461. [Online]. Available: http://bks0.books.google.com.ec/books?id=9vvAlOBfq0kC.

9. R. A. Brooks, "Intelligence without Reason," in Proceedings of the 1991 International Joint Conference on Artificial Intelligence, 1991, pp. 569–595.

10. F. Bellifemine, A. Poggi, and G. Rimassa, "JADE - A FIPA-compliant agent framework," CSELT, Tech. Rep., 1999.

11. I. A. Ferguson, "Touring machines: Autonomous agents with attitudes," Computer, vol. 25, no. 5, pp. 51–55, May 1992. doi:10.1109/2.144395. [Online]. Available: http://dx.doi.org/10.1109/2.144395.

12. M. Woolridge and M. J. Wooldridge, Introduction to Multiagent Systems. New York, NY, USA: JohnWiley & Sons, Inc., 2001. ISBN 047149691X.

13. D. Schmidt, M. Stal, H. Rohnert, and F. Buschman, Pattern-Oriented Software Architecture: Patterns for Concurrent and Networked Objects, ser. Wiley Series in Software Design Patterns. John Wiley & Sons, 2000, vol. 2.

14. F. Baader, D. Calvanese, D. McGuinness, D. Nardi, and P. Patel-Schneider, The Description Logic Handbook: Theory, Implementation and Applications. Cambridge University Press, January 2003. ISBN 0521781760.

15. P. Hitzler, M. Krtzsch, B. Parsia, P. F. Patel-Schneider, and S. Rudolph, "OWL 2 Web Ontology Language Primer," World Wide Web Consortium, W3C Recommendation, October 2009. [Online]. Available: http://www.w3.org/TR/owl2-primer/.

16. Cognitum. Ontorion Semantic Knowledge Management Framework. http://www.cognitum.eu/semantics/ontorion/. Made available on 20 March 2015.

17. B. Quilitz and U. Leser, "Querying distributed rdf data sources with SPARQL," in The Semantic Web: Research and Applications, ser. Lecture Notes in Computer Science, S. Bechhofer, M. Hauswirth, J. Hoffmann, and M. Koubarakis, Eds. Springer Berlin Heidelberg, 2008, vol. 5021, pp. 524–538. ISBN 978-3-540-68233-2. [Online]. Available: http://dx.doi.org/10.1007/978-3-540-68234-9_39.

18. A. Lakshman and P. Malik, "Cassandra: a decentralized structured storage system," Operating Systems Review, vol. 44, no. 2, pp. 35–40, 2010. [Online]. Available: http://dblp.uni-trier.de/db/journals/sigops/sigops44.html#LakshmanM10.

19. P. Kaplanski, "Syntactic modular decomposition of large ontologies with relational database," in ICCCI (SCI Volume), 2009, pp. 65–72.

20. I. Horrocks, O. Kutz, and U. Sattler, "The even more irresistible sroiq." in KR, P. Doherty, J. Mylopoulos, and C. A. Welty, Eds. AAAI Press, 2006. ISBN 978-1-57735-271-6 pp. 57–67.

21. G. Meditskos and N. Bassiliades, "Dlejena: A practical forward chaining owl 2 rl reasoner combining jena and pellet," Web Semantics: Science, Services and Agents on the World Wide Web, vol. 8, no. 1, 2010. [Online]. Available: http://www.websemanticsjournal.org/index.php/ps/article/view/176

22. F. Chang, J. Dean, S. Ghemawat, W. Hsieh, D. Wallach, M. Burrows, T. Chandra, A. Fikes, and R. Gruber, "Bigtable: A distributed storage system for structured data," Proceedings of the 7th USENIX Symposium on Operating Systems Design and Implementation (OSDI06), 2006.

23. T. White, Hadoop: The Definitive Guide, first edition ed., M. Loukides, Ed. O'Reilly, june 2009. [Online]. Available: http://oreilly.com/catalog/9780596521981.

24. L. Lamport, "Paxos made simple, fast, and byzantine," in OPODIS, 2002, pp. 7–9.

25. Cognitum, "Fluent Editor 2014 - Ontology Editor," http://www.cognitum.eu/semantics/FluentEditor/, made available on 20 March 2015.

26. N. E. Fuchs, U. Schwertel, and R. Schwitter, "Attempto controlled english - not just another logic specification language," in LOPSTR 98: Proceedings of the 8th International Workshop on Logic Programming Synthesis and Transformation. London, UK: Springer-Verlag, 1990. ISBN 3-540-65765-7 pp. 1–20.

27. M. Proctor, M. Neale, B. McWhirter, K. Verlaenen, E. Tirelli, A. Bagerman, M. Frandsen, F. Meyer, G. D. Smet, T. Rikkola, S. Williams, and B. Truit, "Drools," 2007. [Online]. Available: http://labs.jboss.com/drools/.

28. OMG. (2008) Semantics of business vocabulary and business rules (sbvr), v1.0. http://www.omg.org/spec/SBVR/1.0/PDF.

29. B. Glimm, M. Horridge, B. Parsia, and P. F. Patel-Schneider, "A syntax for rules in OWL 2." in OWLED, ser. CEUR Workshop Proceedings, R. Hoekstra and P. F. Patel-Schneider, Eds., vol. 529. CEUR-WS.org, 2008. [Online]. Available: http://dblp.uni-trier.de/db/conf/semweb/owled2009.html#GlimmHPP08.

30. M. I. Kellner, R. J. Madachy, and D. M. Raffo, "Software process simulation modeling: Why? what," Journal of Systems and Software, vol. 46, pp. 91–105, 1999.

31. M. A. Mach and M. L. Owoc, "Validation as the integral part of a knowledge management process," in Proceeding of Informing Science Conference, 2001.

Overview of Selected Business Process Semantization Techniques

Krzysztof Kluza, Grzegorz J. Nalepa, Mateusz Ślażyński, Krzysztof Kutt, Edyta Kucharska, Krzysztof Kaczor and Adam Łuszpaj

Abstract Business Process models help to visualize the processes of an organization. There exist several techniques of semantization of Business Processes. We give an overview of Business Process semantization techniques, focusing on the existing approaches in several Business Process Management tools. We also present the use of the existing techniques in the Prosecco (Processes Semantics Collaboration for Companies) research project.

1 Introduction

Business Process Management (BPM) [45] is a modern holistic approach to improving organization's workflow in order to align processes with client needs. BPM focuses on reengineering of processes to obtain optimization of procedures, increase

The chapter is supported by the AGH UST grant. It presents results from the Prosecco project funded by NCBR (The National Centre for Research and Development).

K. Kluza (✉) · G.J. Nalepa · M. Ślażyński · K. Kutt · E. Kucharska · K. Kaczor · A. Łuszpaj
AGH University of Science and Technology, al. A. Mickiewicza 30, 30-059 Krakow, Poland
e-mail: kluza@agh.edu.pl

G.J. Nalepa
e-mail: gjn@agh.edu.pl

M. Ślażyński
e-mail: mslaz@agh.edu.pl

K. Kutt
e-mail: kkutt@agh.edu.pl

E. Kucharska
e-mail: edyta@agh.edu.pl

K. Kaczor
e-mail: kk@agh.edu.pl

A. Łuszpaj
e-mail: adam@softhis.com

© Springer International Publishing AG 2017
T. Pełech-Pilichowski et al. (eds.), *Advances in Business ICT: New Ideas from Ongoing Research*, Studies in Computational Intelligence 658,
DOI 10.1007/978-3-319-47208-9_4

efficiency and effectiveness by constant process improvement. Business Process (BP) models constitute graphical representation of processes in an organization, composed of related tasks that produce a specific service or product for a particular customer [26].

As process models can be ambiguous, some vendors introduced semantization techniques to their BPM solutions. This helps to support more intelligent functions, like web services discovery or element name suggestions. Such techniques often use semantic annotations based on a formally specified ontology. Such an ontology in the simplest form is a unified dictionary of business concepts. Such a dictionary allows for a transparent integration of the elements of business information systems, especially by sharing the conceptualization between different users providing a better decision support facilities.

For over a decade, it has been a common approach to support this task with semantic web technologies, including ontologies. Specifically formalized ontologies e.g. in OWL are of great practical importance. They become not only a tool for capturing the conceptual description of a business system, but also provide a technical backbone for software modules it is composed of. From the technical point of view, building ontologies is a knowledge engineering task that is currently mostly well supported. However, number of challenges remain, including practical integration of dedicated ontologies in a business information systems, often including business process and business rules management modules. Possible unification of these through an ontology is a non trivial task of great importance.

For the purpose of this chapter, we focused on the selected process management systems. We analyze the semantization methods in these solutions. Some of the methods have been presented based on the Prosecco (Processes Semantics Collaboration for Companies) research project,[1] which takes advantage of Activiti, the open-source execution engine. The Prosecco project was a research and development project aims to address the needs and constraints of small and medium enterprises by designing methods that will significantly improve BPM systems.

This chapter is an extended version of the papers [19, 30] presented at the ABICT 2015 workshop. It gives an overview of selected Business Process semantization techniques, extending the discussion of related works.

The chapter is structured as follows. In Sect. 2, we present a short introduction to ontologies. Section 3 gives an overview of business process semantization approaches, especially focusing on the solution developed in the SUPER project, the SAP AG system and the Prosecco project. The chapter is concluded in Sect. 4.

---

[1] See: http://prosecco.agh.edu.pl.

2 Introduction to Ontologies

An ontology is often defined as a formal, explicit specification of a shared conceptualization [39]. In Business Process Management systems, ontologies have to identify key concepts and relations describing various aspects of processes. There are various requirements taken into account by different vendors, e.g.:

- Ontology should be modularized (each module should describe different domain).
- Ontology is designed to be used as concepts dictionary while describing elements of business processes.
- Ontology must be defined in as simple description logic language as possible (OWL Lite$_A$, OWL 2 QL, or eventually OWL 2 RL [16]).
- Each concept and role must be documented with short description (what exactly this term means) and possible connections with other ontology elements.
- There should be a possibility to extend the ontology.

Although there are several methods and tools that support ontology engineering, there are no standardized approaches how to develop ontologies [17]. Brief survey [2] indicates that:

- Existing methods are relatively old.
- The methods can be grouped into categories: incremental and iterative or more comprehensive ones.
- Most methodologies consists of same main steps: assessment, deployment, testing and refinement.
- Most studies suffer from lack of information about tools.
- Only few recent studies suggests decrease in research activity in this field.

There are some well-established methodologies for creating and managing ontologies [2, 17], among them one can distinguished: TOVE (*Toronto Virtual Enterprise*) [14], Enterprise Model Approach [42] and METHONTOLOGY [11].

3 Overview of Business Process Semantization Techniques

This section presents the overview of the semantization approaches for business processes and business process runtime environments, based on the analysis of the research results in this field. It focuses on Business Process semantization in: the SUPER project, the BPM product of the SAP AG company, as well as some substitute of semantization which can be observed in Signavio Process Editor.

3.1   Business Process Semantization in the SUPER Project

The goal of the SUPER project[2] was to create tools for business process semantization by describing process models using concepts from the ontology.

A BPM system that uses ontology as a common language of communication can facilitate clear expressing the statement by the people from the business and provide a method of unambiguous communication between the system, IT people and non-technical people associated with the business.

WSMO Studio is a stand-alone application (also available as a plugin for the Eclipse integrated development environment), which provides the following functions: enables customers to create ontologies, specify goals, web services and mediators, as well as provide for these elements appropriate interfaces. Additionally, the environment provides dedicated editors, including SAWSDL editor to annotate semantics to WSDL.

3.1.1   Ontologies in the SUPER Project

In the SUPER project ontology the following elements can be distinguished (the relationship between them is presented in Fig. 1):

- Web Service Modeling Ontology [37] specifies formally of the terminology of the information used by all other components and provides the semantic description of web services (their functional and non-functional properties, and their interfaces).
- Business Domain Ontologies related to the business domain knowledge (Business Functions, Business Process Resources, Business Roles and Business Modeling Guidelines Ontologies).
- SUPER Ontology Stack:

  - Upper-Level Process Ontology (UPO) is the top-level ontology that aims to represent high-level concepts for Business Process modelling.
  - Business Process Modelling Ontology (BPMO) acts as a bridge between the business level and the processes execution level and is used for representing high-level business process workflows.
  - Semantic Event-driven Process Chains notation Ontology (sEPC) supports the annotation of process models created with EPC tools.
  - Semantic Business Process Modeling Notation Ontology (sBPMN) for formalization of the core subset of the BPMN notation.
  - Semantic BPEL Ontology (sBPEL) extends the ontology of BPEL with a Semantic Web Services model.

---

[2]The website of the SUPER project http://www.ip-super.org/content/view/196/163/ is no longer maintained. However, some pieces of information about the results can be found at: http://www.sti-innsbruck.at/results/movies/sbp-execution-developed-in-super as well as in the project publications.

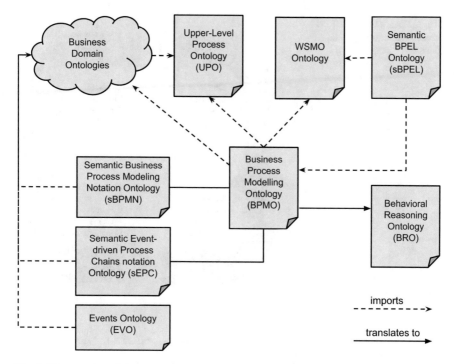

Fig. 1 The overview of the ontologies in the SUPER project (designed based on [12])

- Behavioral Reasoning Ontology (BRO) for reasoning over the business processes behaviours using WSML axioms.
- Events Ontology (EVO) that constitutes a reference model for capturing logging information used by the execution engines and the analysis tools.

3.1.2 The SUPER Project Process Life Cycle

The methodology of the SUPER project defines four phases that form the business process life cycle [44]. For these phases the appropriate methods and techniques for business process semantization were developed [46]. In the following paragraphs, these phases are elaborated (Fig. 2).

Phase 1: Semantic Business Process Modelling

The first phase of the life cycle in the SUPER methodology consists in developing business process models based on the BPMO ontology. It uses the environment for semantic modeling (WSMO Studio tools with the integrated BPMO editor). The business process model is based on the domain ontologies specified for the particular company as well as on Semantic Web Services and Goals. Its source can be implicit

Fig. 2 Phases in the SUPER
project methodology
(designed based on [44])

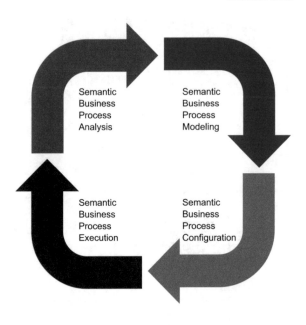

knowledge of business analysts or analysis of reports from the previous Semantic
Business Process (SBP) Analysis phase.

Phase 2: Semantic Business Process Configuration

Semantic Business Process Configuration (SBPC) is a second phase in the SUPER
methodology life cycle that uses the semantic business process models which are
the output from the previous phase. During this phase, the semantic business process
models are configured.

The configuration phase consists in deriving an sBPEL ontology from a BPMO
instance, discovering the possible Semantic Web Services (SWS) [9, 15], identifying
the potential data mismatches, and based on them creating the interface mappings
and data mediators. Lastly, the process is validated in terms of the correctness of the
semantic process description before the execution and potentially refined.

Phase 3: Semantic Business Process Execution

In the third phase of SUPER methodology, modeled and configured processes are
executed and processed. During the runtime, data, which will be used for analysis,
are collected. As this phase is performed without user interactions, it minimizes the
time required for its completion. In this phase, process execution is supported by
the semantic BPEL (BPEL4SWS) and detection and execution of Semantic Web
Services (SWS).

In Fig. 3, the scenario of semantic business process execution is presented. This
scenario involves the following seven steps [44]:

Fig. 3 The execution scenario in the SUPER methodology (designed based on the D10.2 SUPER showcase presentation http://slideplayer.com/slide/741902/)

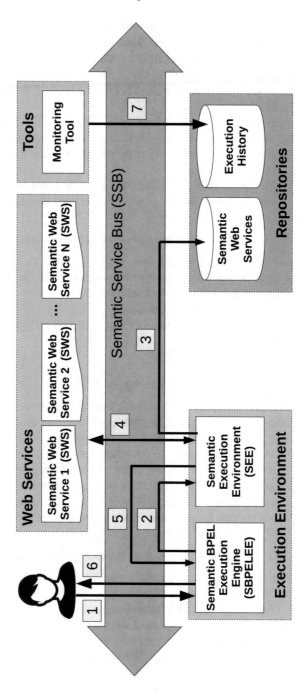

1. *Request Service*—in order to initialize a semantic BPEL process, a user have to send request through the Semantic Service Bus to SBPELEE.
2. *Achieve Goal*—invocation of SWS is delegated to SEE by SBPELEE which passes the WSMO Goal to it.
3. *Discover Service*—SEE queries the Semantic Web Services repository to discover the desired SWS.
4. *Invoke Service*—SEE invokes the discovered SWS.
5. *Engine Return Result*—SEE returns the result received from SWS to SBPELEE.
6. *User Return Result*—After the process execution has been finished, the result is returned to the user.
7. *Process Tracking*—During the execution, execution events are published to Execution History for persistence and to the Monitoring Tool for tracking process executions.

The most important benefits of using such an approach are:

- flexible use of Web Services,
- supplier matching supported by Semantic Web Service discovery and invocation from within semantic business processes,
- more flexible traffic routing,
- automates supplier matching and traffic routing process taking into account all existing suppliers,
- minimizes time-to-offer.

Phase 4: Semantic Business Process Analysis

The last phase of the process life cycle concerns the analysis of the executed processes. In this phase, various analysis goals are supported, such as: overview over process usage, detecting business and technical exceptions, etc.

Moreover, this phase takes advantage of such techniques as Semantic Process Mining or Semantic Reverse Business Engineering. In particular, it provides the AS-IS analysis, exception analysis, as well as user and role analysis. Thanks to this phase, it is possible to get an overall overview about system usage, finding out exceptions within process flow and bottlenecks, as well as get necessary information needed to apply 6 Sigma methodology.

Data, Information, and Process Integration with SWS

The SUPER project took advantage of the experience of the DIP[3] (Data, Information, and Process Integration with Semantic Web Services) [10] platform.

The aim of the DIP platform was extending the semantic web technologies and web services in order to create a new technical infrastructure – Semantic Web Services (SWS). The DIP platform provides: Web Service Modelling Ontology (WSMO), Web Service Modeling Language (WSML) as the language for modeling web services, and Web Service Execution Environment (WSMX) as software framework for runtime binding of service requesters and service providers.

---

[3] See: http://dip.semanticweb.org/.

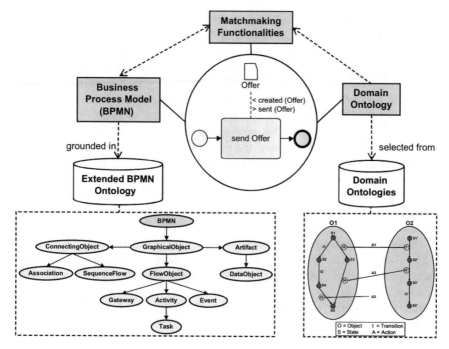

Fig. 4 The main components of the semantic extension tools for modeling business processes in the SAP platform [4]

3.2 Business Process Semantization in SAP AG

Another more extended solution related to business process semantization is an approach developed [4] and patented[4] by the SAP AG company.

Semantization of business processes in the SAP AG uses the semantic descriptions for business process artifacts. The integration consists in linking the identified semantic pieces of information described in the form of ontology with the elements of business process models.

The approach also supports semantic modeling by matching elements of a process model to concepts from ontologies and using fitting functions for choosing proper semantic annotations. This is achieved by comparing the given context and text description with instance domain ontology. In the approach, three goals are achieved: support for modeling, exploring relevant services, and searching the process model repository.

Figure 4 shows the main components of the process modeling tool semantic extension. The BPMN data objects describe activities by defining the related objects and state transitions. For such activities, a user can graphically specify their precondi-

---

[4]See: The patent "Semantic extensions of business process modeling tools" number US 8112257 B2: https://www.google.com/patents/US8112257.

tions and postconditions, as well as define the related objects with the specification of the object state changes before and after the execution of the activity.

An ontology in this approach contains the information about objects, states, state transitions, and actions related to the domain. For each object the possible states and state transitions are defined and they form the object life cycle. These kinds of domain ontologies support semantic process modeling by using their concepts in model elements specification, especially by suggesting relevant concepts or instances of objects. For suggesting the relevant components (data objects, activities, associations and states), a combination of different algorithms associated with the text matching was used. The algorithms take advantage of contextual information related to the process model as well as domain knowledge ontology.

Thanks to the domain knowledge, the names of tasks can be suggested based on the object life cycle. The object life cycle can also be used to exclude re-using the task names that have already been modeled.

The system also supports the semantic description of data flow. The object status can be visualized directly in the diagram. The "less than" sign ($<$) denotes the object status before and the "greater than" ($>$) denotes the object status after performing the associated tasks.

Such semantization supports consistency checking and extends the capabilities of semantic searching. Compared to the approach of the SUPER project, it supports more flexible and accurate semantic annotations by referring directly to the elements from the defined domain ontology.

3.3 Semantization in the Prosecco Project

The *Prosecco* (Processes Semantics Collaboration for Companies) project[5] is a 32 month research and development project funded by NCBR (2012–2015). The motivation of the project was to address the needs and constraints of small and medium enterprises (SME) by designing methods that will significantly improve Business Process Management (BPM) systems. The main goal was to provide technologies that improve and simplify the design and configuration of BPM systems integrated with Business Rules Systems, targeting the management quality and competitiveness improvement. Moreover, fostering decisions making and strategic planning in the SME market sector (mainly in the selected services sector). Specific objectives of the project include:

- development of business process modelling methods taking into account semantic dependencies between business process models and rule models,
- providing recommendation methods for analysis of semantically described business process models, and even more importantly,

---

[5]See: http://prosecco.agh.edu.pl for the project website.

- development of ontology-based mechanisms allowing for creating taxonomies of business logic concepts unifying system objects.

The architecture of the system is oriented towards services what significantly improves its portability and high versatility. Such architecture also enables integration with external tools that provide their functionality as a service. As each element of the architecture provides own data management, there is no centralized repository for all models. Therefore, Prosecco repository consists of several repositories for various models:

1. Repository for Business Processes is called *prosecco-business* and is managed by Activiti engine. It stores information concerning existing processes and their instances, variables and other data processed by the engine. Additionally, the repository contains components that can be used for creating new processes.
2. *Prosecco-knowledgebase* is a repository for rules. It is divided into two parts. The first part stores rules processed by Drools rule engine. The second contains rules learned according to decisions made by the users that are traced on the business process level.
3. Ontology is stored in the OWL format. The POJO model, which is suitable for process and rule engines, can be generated from the OWL representation.
4. *Prosecco-profilemanager* repository stores information related to users and ACL (Access Control List).
5. System History is managed by the Cassandra tool and stores information concerning operations performed within the system.

Due to the fact of usage of the ontology, the project assumes that all data types and their instances existing within the system are consistent with the ontology. Because of the separate data repositories, the object types and existing instances have to be continuously synchronized with the ontology. For example for rule engine the POJO (Plain Old Java Object) model is generated according to the ontology, and for external tools dedicated integration interfaces providing type alignment were developed.

3.3.1 Architecture of the Prosecco System

From the technical point of view, some of the main objectives of the Prosecco system that meets the project goals include the development of the:

1. integrated business logic model composed of business processes and rules,
2. runtime environment suitable for execution of the model,
3. recommendation modules for the design and use of business artifacts,
4. repository of business objects,
5. service bus integrating the system components in a cloud environment, and
6. system ontology based on the taxonomy of shared business concepts.

The outline of the system can be observed in the Fig. 5.

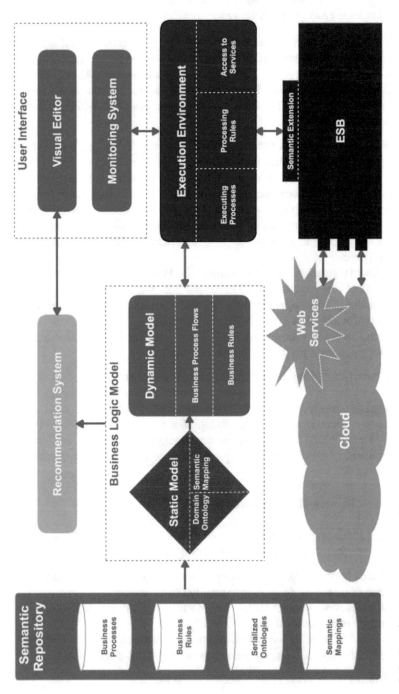

Fig. 5 Outline of the Prosecco system architecture

The end users of the Prosecco system are SMEs, in fact selected employees as well as management of companies. The system aims at supporting carrying out of the main business processes of a company. Moreover, these processes are accompanied by business rules, capturing the details of the business logic, including lower level processing, as well as high-level constraints. Within the system these are modeled with the help of BPMN models, and appropriate business rules models identified with SBVR and implemented with the help of the Drools BRMS [5] as well as the HeaRT rule engine [31].

The process models and rule models are based on concepts captured during initial structured interviews with SMEs. The resulted taxonomy of concepts was used to design and implement the Prosecco ontology that works as the main unifying backbone of the system. It contains all the business terms needed for expressing and capturing the artifacts in business process models and rules. Moreover, it allows to monitor the execution of these models on a semantic level. The execution environment uses the ontology, so users can actually trace how these concepts are used in the executed processes and rules.

From the point of view on the integrators of the system, the ontology supports recommendation mechanism that allow for adaptation of business process and rule models to the needs of specific companies. Furthermore, the repository of business objects uses semantic annotation based on this ontology. It makes it possible for an easy retrieval of needed objects based on semantic queries. In the next section the design of the Prosecco ontology is described.

3.3.2 Ontology in the Prosecco Project

Ontology in the Prosecco project identifies key concepts and relations that describe static aspect of SME sector companies. During Prosecco Ontology development, existing organization ontologies were analysed, such as:

1. *An Organization Ontology* [36]—developed by W3C and Epimorphics Ltd. and implemented in simple description logics language ($\mathcal{SIF}$(D)). It is aimed at describing basic organizational information in a number of domains. Comparing to Prosecco Ontology, this model is less accurate and more general.
2. *IntelLEO Organization Ontology* [18]—created during IntelLEO project and currently not further developed. It models organization structures using people responsibilities and relations between them. This ontology models access rights what is outside the scope of Prosecco Ontology. On the other hand, it is not suitable for describing business processes.
3. *Ontology for Organizations* [40]—is a part of a larger one that is used to annotate The Gazette[6] contents. It is best suitable for characterize activities of the organization (e.g. if it is a government or charity organization). This ontology does not provide concepts for describing organization structure, what was one of main goals of Prosecco Ontology.

---

[6]See: https://www.thegazette.co.uk/.

4. *PROTON* (*PROTo ONtology*) [38]—upper-level ontology that is simple and well-documented. This is the biggest one among analysed models (it consists of 250 classes). As Prosecco Ontology, it describes projects, documents and organization structure. It lacks business processes modelling.

There are also other models which could be used, but because these are not publicly available, there was no possibility to analyse them and compare to the Prosecco Ontology: *O-CREAM-v2, a core reference ontology for the CRM domain* [27]—is a very detailed CRM (Customer Relationship Management) ontology. One of its drawbacks is the lack of emphasis on services; *Unified Enterprise Modelling Ontology* (*UEMO*) [32]—is based on UEML (Unified Enterprise Modeling Language) and depicts companies and information systems. It is coupled with BPM (Business Process Management); *WeCoTin* [41]—ontology designed to modelling process of matching the offer to the customer's requirements.

Prosecco Ontology was partitioned into several parts (see Fig. 6). Each represents a different area in SME management:

1. Project artifacts (PL: *Artefakty*)—components connected with planning and implementation of the project.
2. Organizations (PL: *Organizacje*)—types of companies and their main properties (e.g. e-mail, VAT identification number).
3. Organization structure (PL: *StrukturyOrganizacji*)—elements that describe company structure, e.g. customer care department.
4. Documents (PL: *Dokumenty*)—concepts and relations associated with various kinds of documents.
5. People (PL: *Osoby*)—depiction of people: key properties (e.g. name, surname) and occupation.
6. Methodologies (PL: *Techniki*)—things connected with methods and tools used in company, e.g. code repository.
7. Resources (PL: *Zasoby*)—grouping into human resources, tangible and intangible resources.
8. Events (PL: *Zdarzenia*)—event types and their properties.
9. Prosecco (PL: *Prosecco*)—main module, the parent of other modules that integrates them and adds additional values common to all of them, e.g. uid or name.

Figure 6 depicts all the above parts (modules). Each part consists of at least one concept, object- or data-property from another module. Due to fact that there are lot of relationships between modules, Prosecco module was introduced to integrate all of them (that is why Prosecco module was depicted as the biggest box in Fig. 6).

Although each part is independent of each other (the only links that are between them is really ties 8 modules with the main module Prosecco), there can be found some relationships between them—these are indicated by arrows (the larger the arrow is, the more related the two elements are). The colors indicate similarities between the structure of the ontology.

These nine parts constitute modules of the Prosecco Ontology that consists of: 86 classes, 70 object properties and 64 data properties. All of them are described by 1042

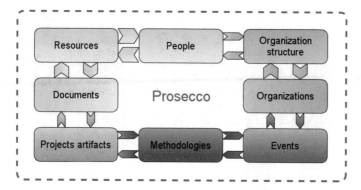

Fig. 6 The parts of the Prosecco Ontology

axioms. Classes are arranged in a hierarchy using `rdfs:subclassof` properties. Besides this simple generalization/specialization relations, each class can be inferred from object and data properties that this class has, e.g. `Project` is something that aggregates some `Tasks` or something that is managed by the `Project leader`. Axioms allows inference only in one way: if something aggregates `Tasks`, it must be a `Project`. Fact that something is a `Project` does not infers conclusion that it must aggregates `Tasks`.

3.4 Other Semantization Methods

There are many other semantization techniques which can be used for semantization of business processes or business process management systems. An overview of various semantic technologies in business process management can be found also in [13], as well as in two papers of the SemTechBPM working group outlining the research directions in business process modeling and analysis [6, 7]. Selected approaches to process semantization methods are briefly summarized in this section.

3.4.1 REST Interface Semantization

As many process execution environments use REST interfaces, the possibility of semantization of the REST interface is presented in [28]. The authors compare the different semantic annotation languages for REST interfaces and show how to take advantage of them by creating a website which combines online applications from different sources (in particular internet services)—the so-called mashup. They also proposed a new language SAWADL, based on WADL (Web Application Description Language), for REST Web Services semantization.

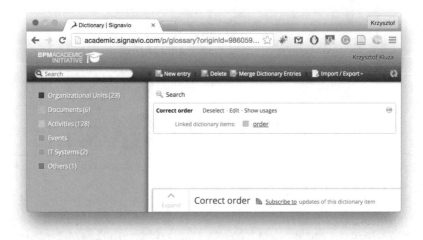

Fig. 7 Using dictionary during modeling in the Signavio Process Editor

3.4.2 Dictionary Supported Modeling

In the Signavio Process Editor,[7] some basic semantization in the form of a dictionary can be observed. In the dictionary one can define the concepts, assign them to one of the 6 categories (Organizational Unit, Documents, Activities, Events, IT System, Other), add the appropriate descriptions, as well as assign to them additional documents (links to them) (see Fig. 7). Then, these concepts can be used to describe the elements of the BPMN model, e.g. during choosing a name for the particular task in the process (see Fig. 8).

Although the tool does not support the formal semantic description in the form of the ontology, it supports multi-lingual description of the same concepts what allows users to work with the same model in different countries in their own languages.

3.4.3 Ontology Support in Process Modeling and Analysis

Other works related to semantic business process modeling can be found in the papers [21, 22], where the processes in the form of Petri nets are connected with the ontology describing the network. Their objective is to standardize the terminology, in particular with regard to the level of abstraction of the labels used in the model. This allows for validating models and detecting the incorrect names of the elements. The ontology for such approaches is mostly developed manually; however, there are approaches in which automated analysis is used [1].

---

[7]See: http://academic.signavio.com.

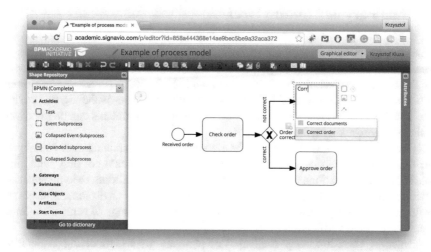

Fig. 8 Using dictionary during modeling in the Signavio Process Editor

Another industrial solution SWB Process uses ontologies to agilely support constant changes in the processes of organizations, in order to increase the degree of automation of the BPM life-cycle [33].

Semantic matching is an important issue in a group of solutions in the area of business process matching. One of the first solutions which was not limited to syntactical similarity assessment, but took advantage of semantic techniques was presented in [23]. This approach consists in generating matching propositions based on automatically annotated activity labels and refining them based on based on behavioral relations of process models and probabilistic optimization. Another solution for automating the task of matching business process models and search for correspondences with regard to the model semantics using Semantic Model Alignment was proposed by Fengel [8]. It extracts the semantic knowledge contained in Business Process models and takes advantage of combination of ontology matching with information linguistics techniques. However, there are also other important issues in matching business process models, like resolving lexical ambiguities in models, e.g. caused by homonyms and synonyms [35]. A broader overview of semantics in process model matching and analysis can be found in [6].

3.4.4 Other Solutions

A bunch of solutions which takes advantage of semantic techniques for model elicitation include semantic process mining, process discovery from text, planning-based process model construction, process model abstraction process maintenance and improvement, process model enrichment and process model translation. The overview of these approaches can be found in [7].

4 Summary and Future Works

The chapter gives an overview of business process semantization approaches in selected business process management systems. We focus on the solutions developed in the SUPER project, the SAP AG system and the Prosecco (Processes Semantics Collaboration for Companies) project. By presenting the state-of-the-art, we provide the insights how the Business Process Management environments can be improved using semantization methods.

Our future works will focus on development of new semantization techniques that can improve business process management environments not only with simple semantic annotations [24, 25]. This can be used for extending recommendation methods in Activiti [3] or semantization of rules in the Wiki environment integrated with processes [20]. Moreover, it is possible to take advantage of semantization in business process verification [43].

However, there are many challenges of semantic process modeling. A detailed overview of the challenges can be found in [29]. Especially semantic interoperability is one of the perspectives for the future systems [34].

Acknowledgments The authors wish to thank the organizers of the ABICT 2015 workshop on the FedCSIS 2015 conference for providing an ample environment for presenting and discussing our research.

References

1. Becker, J., Pfeiffer, D., Falk, T., Räckers, M.: Semantic business process management. In: Handbook on Business Process Management 1, pp. 187–211. Springer (2010)
2. Bergman, M.: A brief survey of ontology development methodologies (2010). http://www.mkbergman.com/906/a-brief-survey-of-ontology-development-methodologies/
3. Bobek, S., Nalepa, G.J., Grodzki, O.: Integration of activity modeller with bayesian network based recommender for business processes. In: G.J. Nalepa, J. Baumeister (eds.) Proceedings of 10th Workshop on Knowledge Engineering and Software Engineering (KESE10) co-located with 21st European Conference on Artificial Intelligence (ECAI 2014), Prague, Czech Republic, August 19 2014, *CEUR Workshop Proceedings*, vol. 1289 (2014). http://ceur-ws.org/Vol-1289/kese10-05_submission_10.pdf
4. Born, M., Dörr, F., Weber, I.: User-friendly semantic annotation in business process modeling. In: M. Weske, M.S. Hacid, C. Godart (eds.) Web Information Systems Engineering – WISE 2007 Workshops, *Lecture Notes in Computer Science*, vol. 4832, pp. 260–271. Springer Berlin Heidelberg (2007). doi:10.1007/978-3-540-77010-7_25
5. Browne, P.: JBoss Drools Business Rules. Packt Publishing (2009)
6. Fellmann, M., Delfmann, P., Koschmider, A., Laue, R., Leopold, H., Schoknecht, A.: Semantic technology in business process modeling and analysis. part 1: Matching, modeling support, correctness and compliance. In: Proceedings of the 6th International Workshop on Enterprise Modelling and Information Systems Architectures (EMISA 2015), September 3–4, 2015. Innsbruck, Austria (2015)
7. Fellmann, M., Delfmann, P., Koschmider, A., Laue, R., Leopold, H., Schoknecht, A.: Semantic technology in business process modeling and analysis. part 2: (semantic) process model elicitation. In: Proceedings of the 6th International Workshop on Enterprise Modelling and

Information Systems Architectures (EMISA 2015), September 3–4, 2015. Innsbruck, Austria (2015)

8. Fengel, J.: Semantic technologies for aligning heterogeneous business process models. Business Process Management Journal **20**(4), 549–570 (2014)

9. Fensel, D., Facca, F.M., Simperl, E., Toma, I.: Semantic web services. Springer Science & Business Media (2011)

10. Fensel, D., Lausen, H., Polleres, A., de Bruijn, J., Stollberg, M., Roman, D., Domingue, J.: Enabling semantic web services: the web service modeling ontology. Springer Science & Business Media (2006)

11. Fernández-López, M., Gómez-Pérez, A., Juristo, N.: Methontology: from ontological art towards ontological engineering. In: Proceedings of the Ontological Engineering AAAI-97 Spring Symposium Series. American Association for Artificial Intelligence (1997)

12. Filipowska, A., Kaczmarek, M., Stein, S.: Semantically annotated EPC within semantic business process management. In: Business Process Management Workshops, pp. 486–497. Springer (2009)

13. Gábor, A., Szabó, Z.: Integration of Practice-Oriented Knowledge Technology: Trends and Prospectives, chap. Semantic Technologies in Business Process Management, pp. 17–28. Springer Berlin Heidelberg, Berlin, Heidelberg (2013)

14. Gruninger, M., Fox, M.S.: The design and evaluation of ontologies for enterprise engineering. In: Workshop on Implemented Ontologies, European Workshop on Artificial Intelligence, Amsterdam, The Netherlands (1994)

15. Hepp, M., Leymann, F., Domingue, J., Wahler, A., Fensel, D.: Semantic business process management: A vision towards using semantic web services for business process management. In: e-Business Engineering, 2005. ICEBE 2005. IEEE International Conference on, pp. 535–540. IEEE (2005)

16. Hitzler, P., Krötzsch, M., Parsia, B., Patel-Schneider, P.F., Rudolph, S.: OWL 2 Web Ontology Language – primer. W3C recommendation, W3C (2009)

17. Jones, D., Bench-Capon, T., Visser, P.: Methodologies for ontology development. In: Proceedings of IT & KNOWS Conference of the 15th IFIP World Computer Congress (1998)

18. Jovanovic, J., Siadaty, M.: IntelLEO organization ontology. Working draft, IntelLEO (2011). http://intelleo.eu/ontologies/organization/spec/

19. Kluza, K., Kaczor, K., Nalepa, G., Slazynski, M.: Opportunities for business process semantization in open-source process execution environments. In: Computer Science and Information Systems (FedCSIS), 2015 Federated Conference on, pp. 1307–1314 (2015)

20. Kluza, K., Kutt, K., Woźniak, M.: SBVRwiki (tool presentation). In: G.J. Nalepa, J. Baumeister (eds.) Proceedings of 10th Workshop on Knowledge Engineering and Software Engineering (KESE10) co-located with 21st European Conference on Artificial Intelligence (ECAI 2014), Prague, Czech Republic, August 19 2014 (2014). http://ceur-ws.org/Vol-1289/

21. Koschmider, A., Blanchard, E.: User assistance for business process model decomposition. In: In First IEEE International Conference on Research Challenges in Information Science, pp. 445–454 (2007)

22. Koschmider, A., Oberweis, A.: Ontology based business process description. In: EMOI-INTEROP (2005)

23. Leopold, H., Niepert, M., Weidlich, M., Mendling, J., Dijkman, R., Stuckenschmidt, H.: Probabilistic optimization of semantic process model matching. In: Business Process Management, pp. 319–334. Springer (2012)

24. Liao, Y., Lezoche, M., Panetto, H., Boudjlida, N.: Semantic annotations for semantic interoperability in a product lifecycle management context. International Journal of Production Research pp. 1–20 (2016)

25. Liao, Y., Lezoche, M., Panetto, H., Boudjlida, N., Loures, E.R.: Semantic annotation for knowledge explicitation in a product lifecycle management context: A survey. Computers in Industry **71**, 24–34 (2015)

26. Lindsay, A., Dawns, D., Lunn, K.: Business processes – attempts to find a definition. Information and Software Technology **45**(15), 1015–1019 (2003). Elsevier

27. Magro, D., Goy, A.: A core reference ontology for the customer relationship domain. Applied Ontology **7**(1), 1–48 (2012)
28. Malki, A., Benslimane, S.M.: Building semantic mashup. In: ICWIT, pp. 40–49 (2012)
29. Mendling, J., Leopold, H., Pittke, F.: 25 challenges of semantic process modeling. International Journal of Information Systems and Software Engineering for Big Companies: IJISEBC **1**(1), 78–94 (2014)
30. Nalepa, G., Slazynski, M., Kutt, K., Kucharska, E., Luszpaj, A.: Unifying business concepts for smes with prosecco ontology. In: Computer Science and Information Systems (FedCSIS), 2015 Federated Conference on, pp. 1321–1326 (2015)
31. Nalepa, G.J.: Architecture of the HeaRT hybrid rule engine. In: L. Rutkowski, [et al.] (eds.) Artificial Intelligence and Soft Computing: 10th International Conference, ICAISC 2010: Zakopane, Poland, June 13–17, 2010, Pt. II, *Lecture Notes in Artificial Intelligence*, vol. 6114, pp. 598–605. Springer (2010)
32. Opdahl, A.L., Berio, G., Harzallah, M., Matulevičius, R.: An ontology for enterprise and information systems modelling. Applied Ontology **7**(1), 49–92 (2012)
33. Pacheco, H., Najera, K., Estrada, H., Solis, J.: Swb process: A business process management system driven by semantic technologies. In: Model-Driven Engineering and Software Development (MODELSWARD), 2014 2nd International Conference on, pp. 525–532 (2014)
34. Panetto, H., Zdravkovic, M., Jardim-Goncalves, R., Romero, D., Cecil, J., Mezgár, I.: New perspectives for the future interoperable enterprise systems. Computers in Industry **79**, 47–63 (2016). Special Issue on Future Perspectives On Next Generation Enterprise Information Systems
35. Pittke, F., Leopold, H., Mendling, J.: Automatic detection and resolution of lexical ambiguity in process models. Software Engineering, IEEE Transactions on **41**(6), 526–544 (2015)
36. Reynolds, D.: The organization ontology. Recommendation, W3C (2014). http://www.w3.org/TR/vocab-org/
37. Roman, D., Keller, U., Lausen, H., de Bruijn, J., Lara, R., Stollberg, M., Polleres, A., Feier, C., Bussler, C., Fensel, D.: Web service modeling ontology. Applied ontology **1**(1), 77–106 (2005)
38. Simov, K., Kiryakov, A., Terziev, I., Manov, D., Damova, M., Petrov, S.: Proton ontology (proto ontology) (2005). http://www.ontotext.com/documents/proton/protontop.ttl
39. Studer, R., Benjamins, V., Fensel, D.: Knowledge engineering: Principles and methods. Data & Knowledge Engineering **25**(1–2), 161–197 (1998)
40. Tennison, J.: The london gazette ontology, organisation module (2008). https://www.thegazette.co.uk/def/organisation.owl
41. Tiihonen, J., Heiskala, M., Anderson, A., Soininen, T.: Wecotin–a practical logic-based sales configurator. AI Communications **26**(1), 99–131 (2013)
42. Uschold, M., King, M.: Towards a methodology for building ontologies. In: IJCAI-95 Workshop on Basic Ontological Issues in Knowledge Sharing, Montreal, Canada (1995)
43. Weber, I., Hoffmann, J., Mendling, J.: Beyond soundness: on the verification of semantic business process models. Distributed and Parallel Databases **27**(3), 271–343 (2010)
44. Weber, I., Hoffmann, J., Mendling, J., Nitzsche, J.: Towards a methodology for semantic business process modeling and configuration. In: E. Di Nitto, M. Ripeanu (eds.) Service-Oriented Computing - ICSOC 2007 Workshops, *Lecture Notes in Computer Science*, vol. 4907, pp. 176–187. Springer Berlin Heidelberg (2009). doi:10.1007/978-3-540-93851-4_18
45. Weske, M.: Business Process Management: Concepts, Languages, Architectures 2nd Edition. Springer (2012)
46. Wetzstein, B., Ma, Z., Filipowska, A., Kaczmarek, M., Bhiri, S., Losada, S., Lopez-Cobo, J.M., Cicurel, L.: Semantic business process management: A lifecycle based requirements analysis. In: Proceedings of the Workshop on Semantic Business Process and Product Lifecycle Management (SBPM 2007), vol. 251, pp. 1–11 (2007)

Selected Approaches Towards Taxonomy
of Business Process Anomalies

Anna Suchenia, Tomasz Potempa, Antoni Ligęza,
Krystian Jobczyk and Krzysztof Kluza

Abstract Modeling based on a graphical notation understandable for different specialists has become very popular. Within the area of business processes, the most common one is the Business Process Modeling and Notation (BPMN). BPMN is aimed at all business users who design, analyze, manage and monitor business processes. BPMN specification is relatively precise, but it provides a descriptive form presented at some abstract, graphical level. The main focus of this chapter is an attempt to present an overview of the anomalies which are likely to occur when modeling with use of BPMN.

1 Introduction

Currently, the approach to modeling based on a graphical notations understandable for different specialists has become very popular. Business process (BP) models are graphical representations of processes in an organization. Business Process Model

A. Suchenia (✉)
Cracow University of Technology, ul. Warszawska 24, 31-155 Kraków, Poland
e-mail: asuchenia@pk.edu.pl

T. Potempa
State Higher Vocational School in Tarnow, 8 Mickiewicza st., 33-100 Tarnow, Poland
e-mail: t_potempa@pwsztar.edu.pl

A. Ligęza · K. Kluza
AGH University of Science and Technology, al. A. Mickiewicza 30,
30-059 Krakow, Poland
e-mail: ligeza@agh.edu.pl

K. Kluza
e-mail: kluza@agh.edu.pl

K. Jobczyk
Laboratory GREYC, University of Caen, Marechal Juin 6, 14032 Caen, France
e-mail: krystian.jobczyk@unicaen.fr

© Springer International Publishing AG 2017
T. Pełech-Pilichowski et al. (eds.), *Advances in Business ICT: New Ideas
from Ongoing Research*, Studies in Computational Intelligence 658,
DOI 10.1007/978-3-319-47208-9_5

and Notation (BPMN)[1] [1, 2] is the most common notation for modeling processes, developed by Business Process Management Initiative and currently supported by the Object Management Group (OMG) because the two organizations merged in 2005. In March 2011, the most recent specification of BPMN (BPMN 2.0) was released. The purpose of BPMN was to create a uniform notation of business processes that would be generally understandable—from professional process analysts, through managers to ordinary workers. According to [2], BPMN 'a standard Business Process Model and Notation (BPMN) will provide businesses with the capability of understanding their internal business procedures in a graphical notation and will give organizations the ability to communicate these procedures in a standard manner. Furthermore, the graphical notation will facilitate the understanding of the performance collaborations and business transactions between the organizations'. The main aims of BPMN include the following:

- process visualization which uses a graphical presentation of a business process. This form of visualization is much more effective than a textual representation;
- documentation through specification of process features;
- communication—provides a set of simple, commonly understandable notations.

BPMN is aimed at all business users, from the analysts, who create the initial process drafts, through the technical developers, whose responsibility it is to implement the technology performing those processes, and finally, to the business people, who will manage and monitor the afore-mentioned processes. The notation is clearly identified by various groups of experts, not only those connected with the IT industry. Yet, in spite of numerous endeavors, problems with unambiguous interpretation still exist. This fact stems from lack of a satisfactory BPMN interpreter. In fact, no formal of BPMN processes was defined, and—as a consequence—no semantics of BPMN components and connection is provided. Hence, various devices can interpret BPMN differently. The fact that there is no formal semantics may lead to misinterpretations and errors.

Majority of papers in the business process area focus on making use of the possibilities that BPMN brings, but papers analyzing errors and ways of eliminating them belong to a minority. BPMN specification is precise but it is only a descriptive, graphical form. Hence, the subject of this chapter forms an attempt to analyze the issue of the anomalies which are likely to occur in BPMN.

This chapter is an extended version of the paper [3] presented at the ABICT 2014 workshop. It is a kind of survey on anomalies in BPMN. An attempt has been materialized at presenting possible problems, both of static, structural and dynamic nature. The research is based on literature analysis and some experience with BPMN models [4–8].

The rest of the chapter is organized as follows. Section 2 covers the presentation of various issues concerning the BPMN notation. In Sect. 3, a literature overview on anomalies is presented. Section 4 presents various types of anomalies in busi-

---

[1] See: http://www.bpmn.org/.

ness processes. It touches potential misinterpretations and errors. The final section contains the summary and conclusions.

2 Business Process Model and Notation

BPMN model consists of simple diagrams made up of a limited set of graphical elements. Simplification of activity flows and processes is clearer for business users and developers. There are four main elements of BPMN, namely: Flow Objects, Connecting Objects, Swimlanes and Artifacts (see Fig. 1).

2.1 BPMN Elements

Flow objects are the key elements describing BPMN. They consist of three core elements: events, activities and gateways [2].

2.1.1 Activities

An Activity is represented by a rounded-corner rectangle and describes a type of work that has to be completed within a business process. There are two kinds of Activities: Tasks and Sub-processes. Task means a single unit of work which is not or cannot by divided in the next stage of business processes specification; in certain sense a task is of an *atomic* nature. On the other hand, sub-process is used for complex work which can be divided into smaller units. It is applied in order to cover or uncover additional specification levels of business processes.

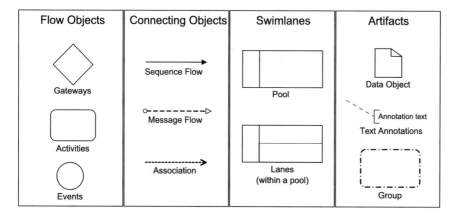

Fig. 1 BPMN core objects of internal Business Process Model

2.1.2   Events

An Event is represented by a circle and means something that happens (compared to an Activity, which is something that has been done). The circular figures differ depending on the type of Event. Events may have an impact on a business process. An event can be an external or internal one. As long as they can influence the process being modeled, they should be modeled.

In general, there are three types of Events: Start, Intermediate and End. Start Event works like a trigger to a process. It is important for every process to have a Start Event to show the beginning of the business process. It allows readers to locate in the BPMN diagram where the process begins, and under what conditions. End event is used to indicate where the business process finishes. It presents the outcome of the process. Intermediate Event represents what happens in the gap between Start Event and End Event. It is responsible for driving a business flow based on the event it specifies (Fig. 2).

2.1.3   Gateways

Gateways are elements used to monitor the way in which some business process flows interact with the others. A Gateway is represented by a diamond shape. Some of the typical types of gateways are the following ones:

- **Data-Based Exclusive Gateway**—it is used to control process flow based on given process data.
- **Inclusive Gateway** can be used to create parallel paths. The conditions of all outgoing flow are evaluated (Fig. 3).
- **Parallel Gateway**—it is used to model the execution of parallel flows without the need of checking any conditions, all outgoing flows must be executed at the same time (Fig. 4).
- **Event-Based Gateway**—it is used to model alternative paths that are based on events (Fig. 5).

2.1.4   Connecting Objects

Flow objects are connected to each other using Connecting objects, which are of three types: sequences, messages, and associations [2]. Sequence Flow is used to show the order in which particular activities will be performed in a process. Message Flow is used to show the flow of messages between two process participants entitled to send and receive them. Association is used to link information and artifacts to activities, events, gateways and flows.

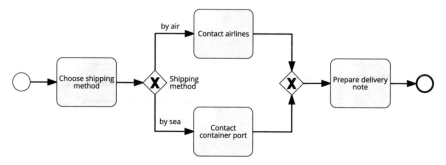

Fig. 2 Data-based exclusive gateway

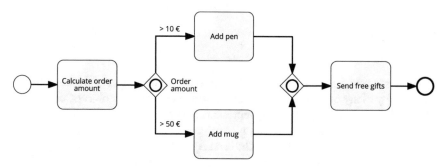

Fig. 3 Inclusive gateway

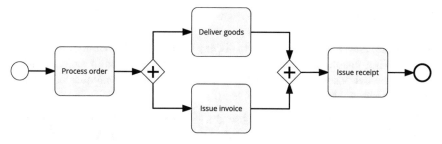

Fig. 4 Parallel gateway

2.1.5 Swimlanes

BPMN usually uses the concept of swimlanes in order to demonstrate what business function a particular activity is connected with or what system executes it. There are two types of swimlane objects: lanes (sub-partition of pools) and pools (represent participants in a business process) [2].

A pool acts as a graphical container for partitioning a set of Activities from other Pools. Lanes can be used to represent specific objects or roles engaged in a process. They are exploited to organize and categorize activities in a pool, according to the

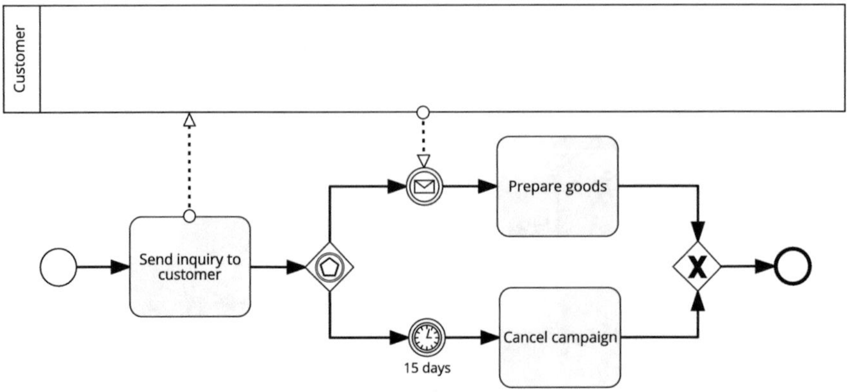

Fig. 5 Event-based gateway

function and role. They are represented by a rectangle extending either vertically or horizontally along the length of the pool. A lane contains flow objects, connecting objects and artifacts. A pool usually represents an organization and a lane constitutes a department within this organization. A process flow can change the lane when different means are needed to fulfill the task. Apart from an organization, a pool can represent other things like functions, applications, localization, classes or units.

2.1.6 Artifacts

Artifacts are diagram elements used to display additional information relative to the process. They enable programmers to include more information in a model. In this way, the model becomes clearer. BPMN does not restrict the number of artifacts, though currently three have been defined [2]:

- *Data objects* are a mechanism whose aim is to show how data is prerequisite or result from activities. They are connected to activities though Associations.
- *Groups* can be used for analysis or documentation objectives but they do not affect the sequence flow.
- *Annotations* are a mechanism used in modeling to provide additional text information for the reader of BPD (Business Process Diagram).

2.2 BPMN Metamodel

Although the BPMN 2.0 specification defines connection rules for sequence and message flows connecting elements, it also provides the metamodel for the notation. Metamodel is one of methods used for defining the syntax of visual languages. The abstract syntax of BPMN is defined by the BPMN metamodel, which is a part of the BPMN 2.0 specification [2]. A fragment of the metamodel defining process and flow elements is presented in Fig. 6.

BPMN 1.0 specification [9] described the semantics of BPMN flow objects in natural language as well as the mapping of BPMN elements to BPEL4WS [10].

One of the fundamental definitions in BPMN 1.0 was the specification of inclusive gateway behaviour [9]: *Process flow shall continue when the signals (tokens) arrive from all of the incoming sequence flow that are expecting a signal based on the upstream structure of the process (e.g. an upstream inclusive decision).* Using this definition, a process modeling the so-called symmetric vicious circle has ambiguous semantics (see Fig. 7), as it is not clear if the process flow may continue.

In the BPMN 2.0 specification [2] some of these ambiguities were clarified. Currently, *the inclusive gateway is activated if at least one incoming sequence flow has at least one token and for each empty incoming sequence flow, there is no token in the graph anywhere upstream of this sequence flow, i.e. there is no directed sequence flow path from a token to this sequence flow unless*

- *the path visits the inclusive gateway, or*

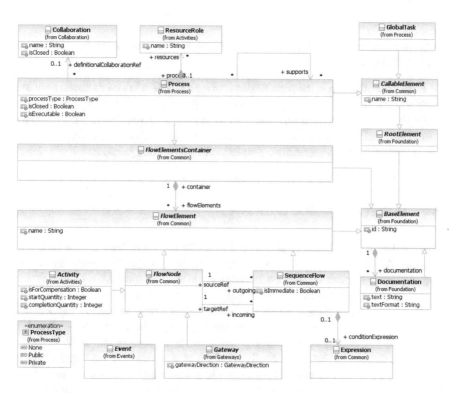

Fig. 6 Fragment of the BPMN MOF metamodel [2]

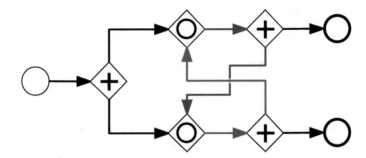

Fig. 7 Symmetric vicious *circle* with inclusive gateways

Table 1 Different exclusive choice representations in BPMN 2.0 [13]

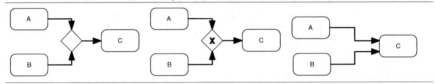

- the path visits a node that has a directed path to a non-empty incoming sequence
  flow of the inclusive gateway.

Such a clarification in the specification solves a problem in simple cases, however
BPMN 2.0 semantics is mostly designed for safe processes e.g. processes that not
suffer from lack of synchronization, etc. Moreover, although the natural language
specification of the semantics is precise enough for intuitive understanding, it is often
not sufficient for implementation, simulation or verification purposes.

2.3 BPMN Equivalent Structures

Two models with different structure, but behaviorally equivalent, can be both correct and unambiguous. This fact stems from the BPMN specification allowing for
expressing the same semantics using various syntactic structures.

This can cause difficulties in modeling or understanding of the model as well as
analysing models. Although behaviorally equivalent structures can be replaceable,
some of them may be not translatable to other languages in order to be analyzed or
verified [8, 11] what generates some practical problems with model analysis.

Such behaviorally (or semantically) equivalent structures can be transformed to
the equivalent model to make it consistent in a way which it might not have been
before [12]. Table 1 presents an example of different exclusive choice representations
in BPMN 2.0 [13].

In this area, there is a need to compare process models [14], there is ongoing research in the area of process models equivalences [13–15] as well as process similarity detection [16–19].

Analysis of Petri nets models equivalences can be found in [14, 15], while the thorough research in the area of BPMN models equivalences was carried out by Vitus Lam and can be found in his papers [12, 20, 21]. Although Lam's equivalences of models are formalized, he analyzes only several equivalence patterns. A wider range of equivalence patterns was presented in the overview paper by Kluza and Kaczor [22].

Similarity measures for business process models have been suggested for different reasons, e.g. measuring compliance between reference and actual models or searching for related models in a repository [16]. Similarity is typically quantified by a distance function that captures the amount of differences between a pair of process models [18]. Respectively, the distance between two process models that are equivalent should be close to zero (process models are identical if their distance equals zero). The extensive research on similarity search and measure was conducted by Dijkman et al. [16–19]. They proposed graph matching algorithms for searching similar process models [17], developed structural similarity measure that compares element labels as well as the topology of process models in order to estimate model similarities [16], as well as developed the technique for fast similarity search that uses similarity estimation based on small characteristics of model fragments for finding potentially relevant models to be compared using the graph matching techniques [19]. There are also solutions which take advantage semantic techniques [23–25]. The overview of semantic-based techniques in process model matching can be found in [26].

2.4 Modeling Rules

BPMN 2.0 defines more than 100 elements, thus practitioners differentiate them based on the degree of model detail. Three levels of models can be distinguished [27]: descriptive level, which is the basic level that uses a very intuitive subset of BPMN to reflect a "happy path" scenario and all major activities in a process; analytical level, dedicated to analysts, modelers and business architects that use complex structures and elements to design fully representative processes, and executable level for technicians in which execution details can be captured in the model.

There are several modeling guidelines that support different techniques and styles. Becker et al. describe the Guidelines of Modeling (GoM) with quality considerations for process models [28]. GoM consists of six guidelines to improve the model quality by taking into account such issues as correctness, relevance, economic efficiency, clarity, comparability, and systematic design.

Mendling et al. focused on the prioritizing several guidelines with the help of industry experts, and then they define desirable characteristics of a process model

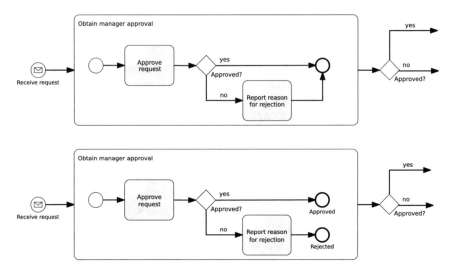

Fig. 8 Exemplary models with correct BPMN syntax but violating (*above*) or following (*below*) a style rule defined by Silver [27]

by formulating seven guidelines which should be taken into consideration when modeling business processes [29]:

1. Model as structured as possible.
2. Decompose a model with more than 50 elements.
3. Use as few elements in the model as possible.
4. Use verb-object activity labels.
5. Minimize the routing paths per element.
6. Use one start and one end event.
7. Avoid OR routing elements.

Both presented guidelines constitute rather a static view by focusing on the resulting process model, but not on the act of modeling itself.

The guidelines concerning modeling style were proposed by Silver [27]. He described 58 style rules which constitute basic principles of compositions intended to make process logic clear. One of the examples of such a rule is illustrated in Fig. 8. The figure presents two models, both of which are correct in terms of the BPMN syntax. However, the model above violates the rule no. 10 defined by Silver in the following way [27]: *If a subprocess is followed by a gateway labeled as a question, the subprocess should have multiple end events, and one of them should match the gateway label.*

A more empirical approach that focuses on the modeler's interactions with the modeling environment was proposed by Pinggera et al. [30]. They investigated the real process followed to create the process model. In their research, they used cluster analysis to identify different modeling styles. As a result, they distinguished three different modeling styles and validated the retrieved clusters using a series of measures for quantifying the process of process modeling [31]. However, they did not specify any guidelines, so their conclusions are not very useful for modelers so far.

Design patterns constitute a practical solution developed in Software Engineering [32]. Currently, they are often applied as general reusable solutions to commonly occurring problems within a given context. The next subsection provides a short overview of workflow patterns which are a specialized form of design patterns that refer to solutions related to process modeling.

Signavio Editor allows for specifying modeling rules which are imposed on model. Modeling rules in Signavio Editor are concerned with (see Fig. 9):

1. architecture,
2. notation,
3. naming,
4. process structure,
5. layout, and
6. consistency.

In the case of architecture, checking includes incorporation of open comments, unique diagram names, and numbering schema in diagram names. As it comes to the notation, a usage of the elements from the defined BPMN subset is checked, as well

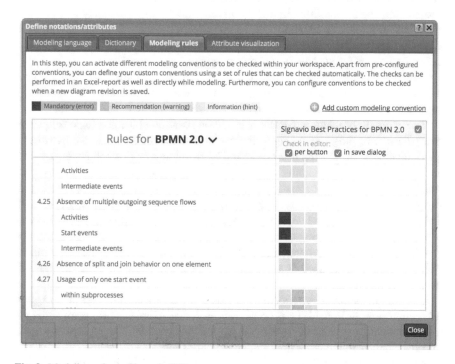

Fig. 9 Modeling rules in Signavio Editor

as mandatory attributes, definition of required dictionary links, and consistency with attributes of the linked dictionary item. The naming consist in checking required element names, consistency of naming style, and uniqueness of element names, In the process structure, usage of different elements in various contexts can be checked, e.g. usage of activities before or-splits, consistent usage of signals correct usage of boundary events, message flows, etc. What is more, absence of loops, deadlocks, multi merges, subprocess relation cycles, multiple incoming sequence flows can be checked. The process layout rules concerns diagram size, colors, edge directions, edge overlays, distances between elements, etc. The last class of rules is concerned with the consistency in the context of decisions: decision logic for decisions and gateways behind decisions.

3 Related Works in the Area of Anomalies

There is a possibility of defining incoherent business logic specification and its interpretation. Even in basic processes anomalies are observed [33]. An improvement is required in the mechanism which provides cohesion in detecting anomalies in business processes [34]. Anomalies have been defined in numerous papers, yet a uniform definition was presented in [35] IEEE standard classification for Software Anomalies and it says: 'Each condition different from the expected is an anomaly'.

In business logic an anomaly can be considered as every negative influence on modeling and models. There is a special kind of anomaly—a defect, which blocks the correct and efficient flow of objects completely.

A taxonomy of anomalies was created on the basis of literature. It concerns the flow control, bases and verification rules of data as well as flow accuracy. The taxonomy can make up a base for classification and research on anomaly possibilities.

The anomaly problem in BPMN is based on searching business logic for particular patterns. In [36], typical controls for anti-patterns are searched for by using a query language for BPMN. It is confirmed by deadlocks or livelock patterns which are used improperly. A similar thing happens in [37] where typical gateway constellations leading to problematic situations in the flow work diagram are presented. A comparable situation occurs in [38] as well, where an 'anomaly pattern' is used. This approach is based on detecting anti-patterns in the data flow. The whole thing is based on time logic using a real model control. By making use of different tools, position [39] is focused on various anomalies which stem from formalism or inadequacy of the tools. Yet another approach is a conception based on UML diagrams in development stages [40].

Control flow anomalies concern problems connected with flow control and gateways conditions [40]. In [41], a problem was presented of control over many semantically identical connections between two work flow elements. This multiplicity complicates changes in the work flow, which is not desired.

Another element of flow control are gateways placed in the modeling center. XOR-gateways with undefined gateway conditions can cause practical problems or even be a reason of an error. A similar thing happens when XOR-gateways conditions do not exclude each other and partly or fully overlap. What happens in flow control in case of lack of synchronization is multiple flow execution. For example, branches and some loop instruction cause such an anomaly [42].

Another situation is a flow deadlock. It is a situation in which the work flow is stopped in the current position of the path and cannot be accomplished. Another lock of flow is known as livelock. In [36] it is called an 'infinite loop'. Flow livelock keeps the operating work flow system in an infinite loop. The reason are bad modeling conditions, which prevent leaving the loop. Both cases—deadlock and livelock are described in [36, 42].

Rule-based anomalies are described in numerous papers [43–46]. They involve mainly two problems connected with base rules. First, Rule-base Consistency are anomalies concerning coherence. Problems result from the set of rules, which have determined conditions but different outcomes at the same time. Rule-base livelocks, also called 'circular rules' [46]. Rule-base livelocks and rule-base deadlocks describe a problem with creation rules, which are dependent on one another although they should not. This type of anomaly suggests that rule-base does not encompass the basic context in which it is used. Coverage anomalies concern the rules in which conditions can be fulfilled by the base context but conclusions are modeled in such a way that no effect will ever be seen. Another type of data flow anomaly is based on [47]. Such anomalies are influenced by those data elements which can be processed by workflow activities.

4 Anomalies in Business Processes

There are various kinds of business process anomalies which can occur while process modeling. One of the classifications distinguish [12]: syntactic anomalies and structural anomalies.

4.1 Syntactic Anomalies in Business Process

This kind of anomaly is not dependent on the data type or tokens in activities. The problem is improper utilization of modeling elements. Analysis of Syntactic Business Process anomalies is important while designing a business process model. In this section examples of syntactic anomalies in business processes will be presented. A division into three groups has been made:

- Incorrect usage of Flow Objects;
- Incorrect usage of Connecting Object;
- Incorrect usage of Swimlanes.

4.2   Incorrect Usage of Flow Object

The anomalies of Incorrect usage of Flow Object result from improper use of the
Event, Activity and Gateway.

Incorrect usage of Activities: Invalid use of Start Event or End Event. The
BPMN specification defines the start and end events as optional. However, their
usage is highly recommended, since each process starts and ends somewhere. Without explicitly using start and end events, a regular BPMN process might look like
the process in Fig. 10. This modeling approach is undesirable and could lead to
misinterpretations.

Depending on application, three anomalies can be distinguished. These are: Activities without Activation, Activities without Termination and Invalid use of Receive
Task. Activities without Termination and Invalid use of Receive Task.

- **Activities without activation**. If an activity is situated on a path that has no start,
  then this is an activity without activation. Even if a start of an event is used.
- **Activities without Termination**. An activity without termination happens when
  the activity cannot be brought to an end. Even if End Event is used.
- **Invalid use of receive task**. Receive Task Element is designed to wait for incoming
  messages from outside users in a business process.

Invalid use of Gateway. There are two groups of anomalies: invalid use of Data-
Based XOR Gateway and invalid use of Event-Based XOR Gateway.

- **Invalid use of Data-Based XOR Gateway**. A data-based XOR Gateway relies
  on the arrival of a data token that has traversed the Process Flow. Data-based XOR
  Gateway must be date-based objects.
- **Event-Based XOR Gateway**. According to BPMN, the event-based gateway cannot be used as a merge gateway. It can only be used as a decision type gateway
  (multiple outputs). It is also possible that none of the awaited events will occur, so
  it is recommended to model also a Timer-Event with represents a Timeout situation. If you are not satisfied given conditions it is incorrect to use of Event-Based
  XOR Gateway.

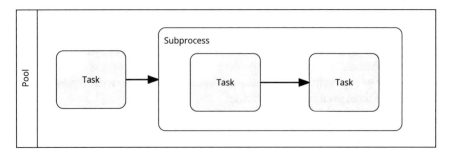

Fig. 10 Implicit process events

Incorrect usage of Connecting Object. Anomalies concerning connecting objects stem from incorrect usage of their elements, that is message flow and sequence flow. As far as incorrect utilization of connecting objects is concerned, a few anomalies can be differentiated: the ones concerning incoming sequence flows, outgoing sequence flows, invalid use of conditional sequence flow. In this case there are two possible irregularities regarding the invalid use of a pool or lane.

4.3   Invalid Use of Pool

When modeling multiple pools, a common mistake is that activities in a Pool are not connected with sequence flows. It is incorrect to use multiple pools as a single process and incorrectly interprets messages flows as way of indicating a sequence of activities (Fig. 11).

Another common problem when modeling multiple pools using tools which allow for incorrect flow modeling is the use of a set of pools as a single pool with multiple lanes. The end result will be an incorrect model (Fig. 12) that represents a single process that spreads over the boundaries of the pool.

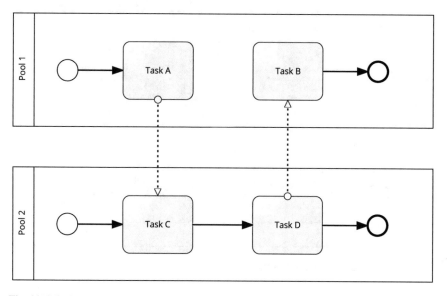

Fig. 11 Missing sequence flow

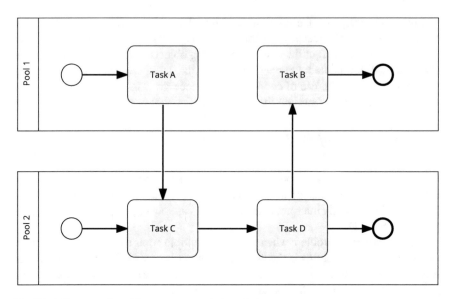

Fig. 12 A Sequence flow May not cross pools boundaries

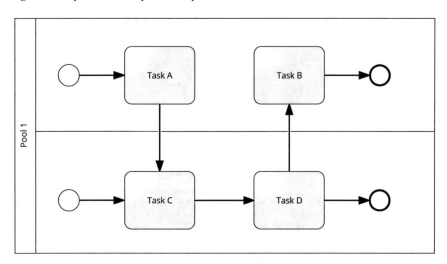

Fig. 13 Two Lances are used as two Pools

4.4 *Invalid Use of Lane*

Improper use of lane as a pool, thereby representing individual processes within separated lanes. This is wrong, because a lane is just a activity-classifying mechanism (Fig. 13).

4.5   Structural Anomalies

Structural anomalies have been described in the literature [48–51]; such anomalies are often classified as one of four types:

- Deadlock;
- Lack of synchronization;
- Dead Activity;
- Infinite Loop.

Note that in fact all the above anomalies correspond to *wrong dynamic behavior*; all of them occur during execution of the process.

A process is sound [52] if and only if it is free of two control-flow errors: the deadlock and the lack of synchronization. First, deadlocks are blockings in the process model, which occur when gateways are used incorrectly. In this case, the links in the process where gateways were installed should be checked. Deadlocks occur when an exclusive gateway was picked for linking and this linkage was combined again with a parallel gateway. They may arise from added intermediate events or multiple exclusive start events, which should be checked again.

A deadlock is a situation where the flow of the Process cannot continue because a requirement of the model is not satisfied. For example, if a Parallel Gateway is expecting a token from all of its incoming Sequence Flow and one never arrives, the process will be trapped with deadlock.

There are two types of deadlocks: deterministic deadlock (Fig. 14) and non-deterministic deadlock (Fig. 15).

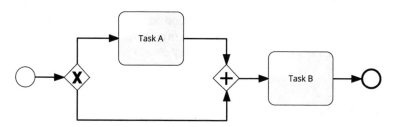

Fig. 14   Deterministic deadlock

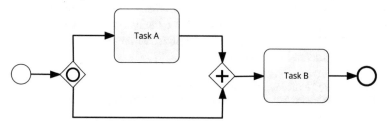

Fig. 15   Non-deterministic deadlock

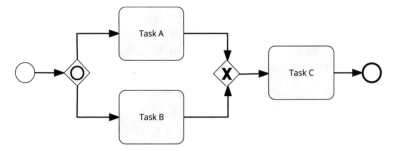

Fig. 16 Lack of synchronization

A deadlock is a reachable state of the process that contains a token on some Sequence Flow that cannot be removed in any possible future. A lack of synchronization (Fig. 16) is a reachable state of the process where there is more than one token on some Sequence Flow. To characterize the lack of synchronization, we follow the intuition that potentially concurrent paths, paths starting with an IOR-split or an AND-split, should not be joined by XOR-join. In the following, we formalize this characterization and show that such structure always leads to lack of synchronization in deadlocks free acyclic workflow graphs [48].

While *Dead Activities* are activities which will never be executed. A last type of anomaly is Infinite Loop, also called 'closed loop'. A closed loop is a cycle without any split. Tokens that enter a closed loop are forever lost to the rest of the workflow. In our model, this leads to a deadlock, because each token entering the closed loop will have a synchronization copy of itself placed on the incoming edge of the initial join that loops back from the cycle. It is hard to imagine a sensible real-world example that contains a closed loop (the BPMN standard document admits this). Banning closed loops from workflows is thus not a serious restriction, especially since infinitely looping cycles are still possible as long as they are not closed [52].

5 Conclusion

BPMN is a popular business process modeling language. The ability of using it is very important in the modeling stage. The main focus of this chapter is an attempt to present an overview of the anomalies which are likely to occur when modeling with use of BPMN. The attempt has been materialized at presenting possible problems, both of static, structural and dynamic nature.

Acknowledgments The authors wish to thank the organizers of the ABICT 2014 workshop on the FedCSIS 2014 conference for providing an ample environment for presenting and discussing our research.

References

1. Allweyer, T.: BPMN 2.0. Introduction to the Standard for Business Process Modeling. BoD, Norderstedt (2010)
2. OMG: Business Process Model and Notation (BPMN): Version 2.0 specification. Tech. Rep. formal/2011-01-03, Object Management Group (2011)
3. Mroczek, A., Ligęza, A.: A note on bpmn analysis. towards a taxonomy of selected potential anomalies. In: Computer Science and Information Systems (FedCSIS), 2014 Federated Conference on, pp. 1097–1102. IEEE (2014)
4. Ligęza, A., Potempa, T.: Artificial intelligence for knowledge management with bpmn and rules. In: Artificial Intelligence for Knowledge Management, pp. 19–37. Springer (2012)
5. Ligęza, A., Potempa, T.: Ai approach to formal analysis of bpmn models: Towards a logical model for bpmn diagrams. In: Advances in Business ICT, pp. 69–88. Springer (2014)
6. Ligęza, A., Kluza, K., Potempa, T.: Ai approach to formal analysis of bpmn models. towards a logical model for bpmn diagrams. In: M. Ganzha, L.A. Maciaszek, M. Paprzycki (eds.) Proceedings of the Federated Conference on Computer Science and Information Systems – FedCSIS 2012, Wroclaw, Poland, 9–12 September 2012, pp. 931–934 (2012). http://ieeexplore.ieee.org/xpls/abs_all.jsp?arnumber=6354394
7. Suchenia, A., Ligęza, A.: Event anomalies in modeling with bpmn. International Journal of Computer Technology & Applications 6(5), 789–797 (2015)
8. Szpyrka, M., Nalepa, G.J., Ligęza, A., Kluza, K.: Proposal of formal verification of selected BPMN models with Alvis modeling language. In: F.M. Brazier, K. Nieuwenhuis, G. Pavlin, M. Warnier, C. Badica (eds.) Intelligent Distributed Computing V. Proceedings of the 5th International Symposium on Intelligent Distributed Computing – IDC 2011, Delft, the Netherlands – October 2011, *Studies in Computational Intelligence*, vol. 382, pp. 249–255. Springer-Verlag (2011). http://www.springerlink.com/content/m181144037q67271/
9. OMG: Business Process Modeling Notation (BPMN) specification. Tech. Rep. dtc/06-02-01, Object Management Group (2006)
10. Sarang, P., Juric, M., Mathew, B.: Business Process Execution Language for Web Services BPEL and BPEL4WS. Packt Publishing (2006)
11. Weidlich, M., Decker, G., Grosskopf, A., Weske, M.: Bpel to bpmn: The myth of a straightforward mapping. In: Proceedings of the OTM 2008 Confederated International Conferences, CoopIS, DOA, GADA, IS, and ODBASE 2008. Part I on On the Move to Meaningful Internet Systems, OTM '08, pp. 265–282. Springer-Verlag, Berlin, Heidelberg (2008)
12. Lam, V.S.W.: Equivalences of BPMN processes. Service Oriented Computing and Applications 3(3), 189–204 (2009)
13. Wohed, P., van der Aalst, W.M.P., Dumas, M., ter Hofstede, A.H.M., Russell, N.: On the suitability of bpmn for business process modelling. In: Business Process Management, 4th International Conference, BPM 2006, Vienna, Austria, September 5–7, 2006, Proceedings, *Lecture Notes in Computer Science*, vol. 4102, pp. 161–176 (2006)
14. van der Aalst, W.M.P., de Medeiros, A.K.A., Weijters, A.J.M.M.: Process equivalence: Comparing two process models based on observed behavior. In: Business Process Management, 4th International Conference, BPM 2006, Vienna, Austria, September 5–7, 2006, Proceedings, *Lecture Notes in Computer Science*, vol. 4102, pp. 129–144 (2006)
15. de Medeiros, A.K.A., van der Aalst, W.M.P., Weijters, A.J.M.M.: Quantifying process equivalence based on observed behavior. Data Knowl. Eng. 64(1), 55–74 (2008)
16. Dijkman, R., Dumas, M., van Dongen, B., Käärik, R., Mendling, J.: Similarity of business process models: Metrics and evaluation. Information Systems 36(2), 498–516 (2011)
17. Dijkman, R., Dumas, M., Garcaa-Banuelos, L.: Graph matching algorithms for business process model similarity search. In: U. Dayal, J. Eder, J. Koehler, H. Reijers (eds.) Business Process Management, *Lecture Notes in Computer Science*, vol. 5701, pp. 48–63. Springer Berlin Heidelberg (2009)

18. Dijkman, R.M., Dongen, B.F., Dumas, M., Garcia-Banuelos, L., Kunze, M., Leopold, H., Mendling, J., Uba, R., Weidlich, M., Weske, M., Yan, Z.: A short survey on process model similarity. In: J. Bubenko, J. Krogstie, O. Pastor, B. Pernici, C. Rolland, A. Solvberg (eds.) Seminal Contributions to Information Systems Engineering, pp. 421–427. Springer Berlin Heidelberg (2013)
19. Yan, Z., Dijkman, R., Grefen, P.: Fast business process similarity search. Distributed and Parallel Databases **30**(2), 105–144 (2012)
20. Lam, V.S.W.: Formal analysis of BPMN models: a NuSMV-based approach. International Journal of Software Engineering and Knowledge Engineering **20**(7), 987–1023 (2010)
21. Lam, V.S.W.: Foundation for equivalences of BPMN models. Theoretical and Applied Informatics **24**(1), 33–66 (2012)
22. Kluza, K., Kaczor, K.: Overview of BPMN model equivalences: towards normalization of BPMN diagrams. In: J. Canadas, G.J. Nalepa, J. Baumeister (eds.) 8th Workshop on Knowledge Engineering and Software Engineering (KESE2012) at the at the biennial European Conference on Artificial Intelligence (ECAI 2012): August 28, 2012, Montpellier, France, pp. 38–45 (2012). http://ceur-ws.org/Vol-949/
23. Fengel, J.: Semantic technologies for aligning heterogeneous business process models. Business Process Management Journal **20**(4), 549–570 (2014)
24. Leopold, H., Niepert, M., Weidlich, M., Mendling, J., Dijkman, R., Stuckenschmidt, H.: Probabilistic optimization of semantic process model matching. In: Business Process Management, pp. 319–334. Springer (2012)
25. Pittke, F., Leopold, H., Mendling, J.: Automatic detection and resolution of lexical ambiguity in process models. Software Engineering, IEEE Transactions on **41**(6), 526–544 (2015)
26. Fellmann, M., Delfmann, P., Koschmider, A., Laue, R., Leopold, H., Schoknecht, A.: Semantic technology in business process modeling and analysis. part 1: Matching, modeling support, correctness and compliance. In: Proceedings of the 6th International Workshop on Enterprise Modelling and Information Systems Architectures (EMISA 2015), September 3–4, 2015. Innsbruck, Austria (2015)
27. Silver, B.: BPMN Method and Style. Cody-Cassidy Press (2009)
28. Becker, J., Rosemann, M., Uthmann, C.: Guidelines of business process modeling. In: W.M.P. van der Aalst, J. Desel, A. Oberweis (eds.) Business Process Management, *Lecture Notes in Computer Science*, vol. 1806, pp. 30–49. Springer Berlin Heidelberg (2000)
29. Mendling, J., Reijers, H.A., van der Aalst, W.M.P.: Seven process modeling guidelines (7pmg). Information & Software Technology **52**(2), 127–136 (2010)
30. Pinggera, J., Soffer, P., Zugal, S., Weber, B., Weidlich, M., Fahland, D., Reijers, H., Mendling, J.: Modeling styles in business process modeling. In: I. Bider, T. Halpin, J. Krogstie, S. Nurcan, E. Proper, R. Schmidt, P. Soffer, S. Wrycza (eds.) Enterprise, Business-Process and Information Systems Modeling, *Lecture Notes in Business Information Processing*, vol. 113, pp. 151–166. Springer Berlin Heidelberg (2012)
31. Pinggera, J., Soffer, P., Fahland, D., Weidlich, M., Zugal, S., Weber, B., Reijers, H., Mendling, J.: Styles in business process modeling: an exploration and a model. Software & Systems Modeling pp. 1–26 (2013)
32. Sommerville, I.: Software Engineering, 7th edn. International Computer Science. Pearson Education Limited (2004)
33. Mendling, J., Verbeek, H., van Dongen, B.F., van der Aalst, W.M., Neumann, G.: Detection and prediction of errors in epcs of the sap reference model. Data & Knowledge Engineering **64**(1), 312–329 (2008)
34. Hallerbach, A., Bauer, T., Reichert, M.: Capturing variability in business process models: the provop approach. Journal of Software Maintenance and Evolution: Research and Practice **22**(6-7), 519–546 (2010)
35. Group, I., et al.: 1044-2009-ieee standard classification for software anomalies. IEEE, New York (2010). https://standards.ieee.org/findstds/standard/1044-2009.html
36. Laue, R., Awad, A.: Visualization of business process modeling anti patterns. Electronic Communications of the EASST **25** (2009)

37. Kühne, S., Kern, H., Gruhn, V., Laue, R.: Business process modeling with continuous validation. Journal of Software Maintenance and Evolution: Research and Practice **22**(6-7), 547–566 (2010)
38. Trčka, N., Van der Aalst, W.M., Sidorova, N.: Data-flow anti-patterns: Discovering data-flow errors in workflows. In: Advanced Information Systems Engineering, pp. 425–439. Springer (2009)
39. Lohmann, N., Wolf, K.: How to implement a theory of correctness in the area of business processes and services. In: Business Process Management, pp. 61–77. Springer (2010)
40. White, S.A.: Process modeling notations and workflow patterns. Workflow handbook **2004**, 265–294 (2004)
41. Olkhovich, L.: Semi-automatic business process performance optimization based on redundant control flow detection. In: Telecommunications, 2006. AICT-ICIW'06. International Conference on Internet and Web Applications and Services/Advanced International Conference on, pp. 146–146. IEEE (2006)
42. Liu, R., Kumar, A.: An analysis and taxonomy of unstructured workflows. In: Business Process Management, pp. 268–284. Springer (2005)
43. Dohring, M., Heublein, S.: Anomalies in rule-adapted workflows-a taxonomy and solutions for vbpmn. In: Software Maintenance and Reengineering (CSMR), 2012 16th European Conference on, pp. 117–126. IEEE (2012)
44. Ligęza, A., Nalepa, G.J.: A study of methodological issues in design and development of rule-based systems: proposal of a new approach. Wiley Interdisciplinary Reviews: Data Mining and Knowledge Discovery **1**(2), 117–137 (2011)
45. Xu, D., Xia, K., Zhang, D., Zhang, H.: Model checking the inconsistency and circularity in rule-based expert systems. Computer and Information Science **2**(1), 12 (2009)
46. Zaidi, A.K., Levis, A.H.: Validation and verification of decision making rules. Automatica **33**(2), 155–169 (1997)
47. Awad, A., Decker, G., Lohmann, N.: Diagnosing and repairing data anomalies in process models. In: Business Process Management Workshops, pp. 5–16. Springer (2009)
48. van der Aalst, W.M., Hirnschall, A., Verbeek, H.: An alternative way to analyze workflow graphs. In: Advanced Information Systems Engineering, pp. 535–552. Springer (2002)
49. Hong, L., Bo, Z.J.: Research on workflow process structure verification. In: e-Business Engineering, 2005. ICEBE 2005. IEEE International Conference on, pp. 158–165. IEEE (2005)
50. Kim, G.W., Lee, J.H., Son, J.H.: Classification and analyses of business process anomalies. In: Communication Software and Networks, 2009. ICCSN'09. International Conference on, pp. 433–437. IEEE (2009)
51. Lin, H., Zhao, Z., Li, H., Chen, Z.: A novel graph reduction algorithm to identify structural conflicts. In: System Sciences, 2002. HICSS. Proceedings of the 35th Annual Hawaii International Conference on, pp. 10–pp. IEEE (2002)
52. Börger, E., Sörensen, O., Thalheim, B.: On defining the behavior of or-joins in business process models. Journal of Universal Computer Science **15**(1), 3–32 (2009)

Hybrid Framework for Investment Project Portfolio Selection

Bogdan Rębiasz, Bartłomiej Gaweł and Iwona Skalna

Abstract Project selection is a complex multi-criteria decision making process with multiple and often conflicting objectives. The complexity of the project selection problems stems primarily from the large number of projects from among which an appropriate collection (an effective portfolio) of investment projects must be selected. This paper presents a new hybrid framework for construction of an effective portfolio of investment projects. The parameters of the considered model are described using both probability distributions and fuzzy numbers (possibility distributions). The proposed framework enables to take into account stochastic dependencies between model parameters and economic dependencies between projects. As a result, a set of Pareto optimal solutions is obtained. The framework is adapted for enterprises with multistage production cycle, i.e., for enterprises of a metallurgical or chemical industry. The performance of the proposed method is illustrated using an example from metallurgical industry.

1 Introduction

Decisions of which investment projects need to be taken up that will provide maximum profit and recognition are key decisions that impact the long-term efficiency and market position of a company. Limited resources cause that not all the investment projects can be undertaken at once. So, a company must select the best projects in order to advance the organizational goals.

B. Rębiasz (✉) · B. Gaweł · I. Skalna
AGH University of Science and Technology, Ul Gramatyka 10, Kraków, Poland
e-mail: brebiasz@zarz.agh.edu.pl

B. Gaweł
e-mail: bgawel@zarz.agh.edu.pl

I. Skalna
e-mail: iskalna@zarz.agh.edu.pl

© Springer International Publishing AG 2017
T. Pełech-Pilichowski et al. (eds.), *Advances in Business ICT: New Ideas from Ongoing Research*, Studies in Computational Intelligence 658,
DOI 10.1007/978-3-319-47208-9_6

There are various methods for selection of a portfolio of projects. Most of them are rarely used in practice, because of the complexity and higher requirements on input data. The usage of some methods is limited by the fact that they inadequately model the risk and uncertainty and do not take into account interrelationships between projects, criteria or parameters. Sometimes, they are too difficult to be understood, and thus, to be used by managers. Whereas, pervasive competitiveness and inevitable uncertainty that accompanies each business activity force managers to utilize modern techniques and tools in the process of selection of the best projects. Additionally, although there are plenty of techniques for managing project portfolio selection, there is a total lack of frameworks for investment projects.

An investment project is a capital used to purchase fixed assets, such as land, machinery or buildings. The process of estimation and selection of investment projects is often named in literature as "capital budgeting". An effective capital budget (a portfolio of investments) is a budget which provides the maximum NPV (*Net Present Value*) for an acceptable level of risk or the lowest level of risk for a given acceptable NPV of a portfolio.

The key stages in the process of selecting investment projects are the choice of an acceptable level of risk, uncertainty treatment and the analysis of dependencies between projects and parameters.

The choice of an appropriate method for risk assessment involves, among others, the description of uncertainty, which is inherent in any business activity. For many years, probabilistic approach was considered as the only appropriate mathematical tool to describe and deal with uncertainty. In fact, it is still the most commonly used in practice and prevails in the literature concerning the risk in business activity. However, last decades have shown that in real problems of assessing risk in business activity, not only randomness, but also imprecise or incomplete data is an important source of information. For this reason, alternative ways of modeling uncertainty, such as fuzzy sets (possibility distributions) or interval numbers, are used increasingly often. The common practice is to unify different ways of representing uncertainty in a single modeling framework. To the best knowledge of the authors, the most appropriate approach to assessing the risk in the selection of investment portfolio is to develop and use methods that allow different representations of uncertainty (e.g., by probability distributions, fuzzy numbers and intervals) to be processed according to their nature and only finally combine them into a synthetic easy-to-interpret risk measure.

Another important aspect of project portfolio selection is the dependency, the one between parameters and the one between projects. The dependency between parameters is usually described statistically and, therefore, is called "statistical dependency". It is used to measure the strength of dependencies (associations) between two random or fuzzy variables (model parameters). Statistical dependency is typically modeled by fuzzy or probabilistic correlation or regression. The dependency between projects (interdependency), usually called "economic dependency", is used to describe interactions between investment projects. Interdependency within a portfolio of investment projects stem from multiple interrelated factors, such as resource and time constraints, financial costs, project outcome and risk profiles. Interdependency is especially challenging to model, due to the difficulties with their description.

This paper briefly presents a novel framework for the selection of investment project portfolio. The proposed framework integrates a non-linear programming with tools that enable to describe interdependency between projects in a situation when model parameters are described both using probability distributions and possibility distributions. The paper is organized as follows. Section 2 outlines different ways of description of uncertainty. In Sect. 3, the current state of art in the selection of a portfolio of investment project is presented. A novel framework for the selection of an effective portfolio of investment projects is suggested in Sect. 4. To demonstrate the effectiveness of the proposed framework, a numerical example is presented in Sect. 5.

2 Description of Uncertainty in the Assessment of Investment Projects

There are many ways to describe uncertainty of input parameters. Probability theory is useful for prediction of events, based on partial knowledge. Fuzzy approach, on the other hand, quantifies the degree of truth. The last decades have shown that the complexity of dependencies, both inside and outside a company, cause that probability theory by itself is no longer sufficient to represent all kinds of uncertainty that appears in the assessment of investment projects. Real-world investment projects usually contain a mixture of quantitative and qualitative data.

In the case of economic calculus, for some parameters it is possible to determine probability distributions (statistical data), whereas for others, the available information is in the form of possibility degrees (subjective assessments of phenomena made by experts [1–3]), so the available data is heterogeneous in nature. To sum up, one may say, after Baudrit et al. [3], that randomness and imprecise or missing information are two reasons of uncertainty. Therefore, in the process of the evaluation of investment projects (estimation of efficiency and risk of projects), it is inevitable to deal with uncertainty caused by vagueness intrinsic to human knowledge and imprecision or incompleteness resulting from the limit of human knowledge [4, 5]. Hence, it is necessary to use a scheme for representing and processing vague, imprecise, and incomplete information in conjunction with precise and statistical data [3, 5, 6].

There are hardly a few studies which describe the use of hybrid data [7–11], i.e., data partially described by probability distributions, and partially by possibility distributions. The use of such data allows to reflect more properly the knowledge on parameters of economic calculus. However, very often, in the assessment of efficiency of investment projects, no distinction is made between these two types of uncertainty, both being represented by means of a probability distribution [1, 3, 5, 6, 12]. Whereas, Ferson and Ginzburg [12] indicate that distinct methods are needed to adequately represent random variability (often referred to as "objective uncertainty") and imprecision (often referred to as "subjective uncertainty").

3   Methods for the Selection of Effective Portfolios of Investment Projects

The problem of capital budgeting was for the first time formulated by Lorie and Savage [4]. Later on, the problem was solved using mathematical programming methods. First works on this subject date back to 1960s and 70s [13–16]. The problem of determining the capital budget was also solved using linear programming, linear programming with binary variables and multi-objective programming methods.

A lot of attention, especially in the recent years, is given to the risk of investment projects. A method for the construction of an effective portfolio of investment projects on the capital market was first presented by Markowitz [17]. Seitz has adopted the ideas of Markowitz for capital budgeting [18]. The significant problem associated with the selection of investment projects that is faced by the Markowitz's model is the indivisibility of assets. Seitz has proposed to solve this problem by using the binary quadratic programming.

Methods for the selection of an effective portfolio of investment projects are being constantly improved [14, 19–22]. Probability distributions of selected parameters were used to describe the uncertainties in these models.

Literature (see, e.g., [23–25]) presents methods for the selection of a portfolio of investment projects when uncertain parameters of efficiency calculus are described by means of fuzzy numbers. A capital budget is modeled there so that efficiency indicators are achieved at the assumed level of credibility or possibility. Kahraman [26] uses dynamic programming for determining the capital budget of a company in fuzzy conditions.

Guyonnet et al. [5] have proposed a method which facilitates estimation of the risk when both probability and possibility distributions are used simultaneously. This method was a modification of the method proposed previously by Cooper et al. [6]. Methods for processing hybrid data combine stochastic simulation with arithmetic of fuzzy numbers. As a result of processing of such data, Guyonnet et al. [5] define two cumulative distribution functions: optimistic and pessimistic. Similarly, Baudrit et al. [3] use probability and possibility distributions in risk analysis. Also in this case, the procedure for data processing combines stochastic simulation with arithmetic of fuzzy numbers. As a result of processing of such data, authors obtain random fuzzy variable, which characterizes the examined phenomenon.

The main deficiency of all the above-described methods is that they do not take into account economic dependencies between projects. Dickinson et al. [27] presented a method for optimal scheduling of investment projects, which takes into account the fact that particular projects can be complementary or substitutive to each other. They used nonlinear programming to select projects. Santhanam and Kyparisis [28] presented a mathematical model for the selection of a portfolio from economically dependent investment projects associated with the development of information systems. They used binary programming for the projects selection. Zuluaga et al. [29] presented a model that enables the selection and scheduling of economically dependent investment projects. However, the models of Dickson, Santhanam and

Kyparisis and Zuluga do not take into account uncertainty of cash flows generated by investment projects and stochastic dependencies between projects. Medaglia et al. [30] proposed the usage of evolutionary algorithms for the selection of economically and stochastically dependent investment projects.

It must be, however, highlighted that there are no methods for the selection of effective investment portfolio, which can process hybrid data, i.e., data expressed in the form of fuzzy numbers and probability distributions. Most of the existing approaches usually unify different ways of uncertainty representation by transforming one form of uncertainty into another. Obviously, such transformation entails some problems. For example, transformation of a probability distribution into a possibility distribution causes the loss of information, whereas the opposite one requires additional information to be introduced. This leads to systematic errors in the estimation of efficiency. It is, therefore, necessary to elaborate a framework for representing and processing stochastic, vague, imprecise, and incomplete information in conjunction with precise data for selection of investment project portfolio. Such a framework should also be able to take into account stochastic and economic dependencies. the main objective of this work is to develop such framework.

4 A Framework for the Selection of Investment Project

The process of building an effective capital budget consists of three phases [31]: strategic consideration, individual project evaluation and portfolio selection. Techniques used in the first phase can assist in the determination of strategic focus and budget allocation for the investment portfolio. In the second phase, projects are evaluated independently in terms of project's individual contribution to portfolio objectives. The third phase deals with the selection of portfolios based on candidate project parameters, including their interactions with other projects through resource constraints or other interdependencies.

Because the approach proposed here focuses on interdependency between projects, the problem of building an effective capital budget is divided into two models— *portfolio selection model* (PSM) and *portfolio evaluation model* (PEM). The purpose of first model is to find selection of the investment projects to gain the best evaluation parameters. Second model is used to determine evaluation parameters for a given set of investment projects.

4.1 Projects Interdependency in Uncertain Environment

PSM focuses on the selection of projects. Most of the project portfolio optimization methods and tools treat each project in portfolio as an isolated entity. This leads to systematic errors in the estimation of risk and efficiency, and usually produces large overestimation. In order to eliminate this deficiency, the interdependency between project should be considered. Three types of projects interdependencies are recognized in the literature: benefit, resource and technical [32–34]. They are briefly described below.

Resources interdependency occurs when the demand for resources to develop projects independently is greater than amount of resources required when all of projects are selected. For example, in order to finish project A, a costly lab equipment is needed. The same equipment is needed to finish project B. These lab equipment may be shared between projects, but since all projects have been treated independently, the cost of lab equipment will be included in the budget of each project. So, the total cost is overestimated by not taking into account the synergy effect.

Benefit interdependency occurs when the total advantage of at least two independent projects increases or decreases when these projects are treated as interrelated. Investment projects may therefore be benefit independent (when no such effect exists) or benefit dependent (when such an effect exists). There are two types of benefit interdependency: complementary and competitive (substitutive). Two projects are complementary when their cumulative benefit is greater then the sum of benefits of each individual project separately. In the extreme case for mutually dependent projects, all are accepted for implementation or none. When benefits of two projects are reduced, then they are called substitutive. So, complementary projects generate the synergy of benefits and competitive cannibalize each other.

Technical interdependency occurs when there is a set of exclusive projects such that only one of them may be selected.

In [35] a degree of interdependence is introduced to model interdependence among investment projects. This metric is used to determine mutual influence of a pair of projects. The values of the metric come from experts judgments. But, in the problem of capital budgeting, due to the complex and dynamic nature of the economic environment and dependency between projects, it is hard to accurately predict the benefits of a given portfolio of projects. Therefore, it is assumed that projects interdependency reflects the dependency between projects' parameters. In the framework proposed in this paper, a mathematical model of an enterprise is built, which reflects all kind of interdependency described above. Then, the efficiency of every portfolio of projects is verified. The main drawback of this approach is that it leads to a combinatorial explosion. To solve the proposed optimization problem with larger number of decision variables, some metaheuristic should be used.

There are five classes of project portfolio selection models [16]: ad hoc approaches (e.g., profiles), comparative approaches [36] (e.g., AHP), scoring models, portfolio matrices, and optimization models. Multiple and often conflicting objectives and interdependency cause that only the latter models can used. These models are based on some form of mathematical programming to support optimization process and to include project interactions such as resource and technical dependencies. PSM is multi-criteria linear programming model, and PEM is non-linear programming model combined with stochastic simulation.

4.2 Portfolio Selection Model (PSM)

Let us consider a manufacturer which plans to build a new departments or extend the existing ones within the steps of primary production process to meet future demand. The management prepared set of possible investment projects that could be realized in the planning horizon consists of h planning periods $T = \{0, 1, \ldots, h\}$. Set W contains indices of those potential projects. Each potential project w is characterized by set of investment outlays η_w^t in each period of its lifetime $t = \{1, \ldots, t_w\}$.

Due to the capital limits, the company should choose only some of potential projects. Let η^τ defines limit of investment outlays in the $\tau \in T$ planning period. The aim of portfolio selection model is to find such combination of potential projects and their start dates to maximize expected financial evaluation parameter and minimize risk. The variable p_w^τ equals one, if project w is realized in τ period. Function $fin(p_w^\tau)$ denote financial evaluation parameter for a given portfolio of investments. There are number of ways to evaluate costs and benefits of investment project. The most commonly used are: payback period, internal rate of return and net present value. Typically the value of $fin(p_w^\tau)$ is uncertainty distribution which reflects uncertainty regarding the projects' parameters. Then, the problem of selection of the portfolio of investments is defined as follows: find p_w^τ that maximizes of the expected value of $fin(p_w^\tau)$ and minimizes $\sigma(fin(p_w^\tau))$ subject to:

1. project-to-period assignment constraints

   - each project is assigned to subset of consecutive planning periods

$$s_w = \begin{cases} \tau \text{ for } y_w^\tau = 1 \\ 0 \text{ otherwise} \end{cases} \tag{1}$$

$$p_w^{\lfloor (\tau_1 + \tau_2)/2 \rfloor} \geq p_w^{\tau_1} + p_w^{\tau_2} - 1; \; w \in W, \tau_1, \tau_2 \in T : s_w \leq \tau_1 \leq \tau_2 \leq h \tag{2}$$

2. capacity constraints

   - each project is assigned to at least as many periods as its lifetime

$$\sum_{\tau \in T} p_w^\tau \geq t_w; \; w \in W \tag{3}$$

3. project allocation constraints

   - in every period the demand on investment outlays cannot exceed investment outlays for this period

$$\sum_{\tau \leq \bar{\tau}} p_w^\tau = y_w^{\bar{\tau}}; \; w \in W, \bar{\tau} \in T \tag{4}$$

$$\sum_{w \in W} \eta_{w, y_w^\tau} \leq \eta^\tau, \tau \in T, \tag{5}$$

Table 1 Notation: PSM model

Indices
w = investment project, $w \in W = \{1, \dots, m\}$
τ = planning period $\tau \in T = \{1, \dots, h\}$
t = period of w investment realization $t \in T_w = \{1, \dots, t_w\}$
Parameters
η^τ = limit of investment outlays in the τ year
h = planning period
m = number of potential investment projects
d_w = duration of w investment project
Variables
p_w^τ = equals 1, if project w is realized in τ period, 0 otherwise
Auxiliary variables
y_w^τ = period of w investment lifetime at τ planning period 0 means investment is not started or finish
s_w = the earliest period at which investment operates
Functions
$fin(p_w^\tau)$ = denote financial evaluation parameter for a given portfolio of investments

4. non-negativity and integrality conditions

$$p_w^\tau \in \{0, 1\}; w \in W, \tau \in T \tag{6}$$

Notation for PSM is summarize in Table 1.

In order to solve PSM problem, the Portfolio Evaluation Model (PEM) to obtain uncertainty distribution of $fin(p_w^\tau)$ was build.

4.3 Portfolio Evaluation Model (PEM)

PEM model is used to computes $fin()$ for a given values of $p_{w,\tau}$ using hybrid method which combines Monte Carlo simulation and non-linear programming. This is a mathematical model of an enterprise with dependency and uncertainty. it may be divided into three main parts which model:

- technical and resources interdependency,
- uncertainty of model parameters and
- statistical dependency between parameters.

Notation used in PEM was presented in Table 2.

Table 2 Notation: PEM model

Indices
w = investment project, $w \in W = \{1, \ldots, m\}$
τ = planning period $\tau \in T = \{1, \ldots, h\}$
i = product $i \in I = \{1, \ldots, q\}$
j = processing stage $j \in J = \{1, \ldots, n\}$
k = parallel departments on stage j, $k \in K_j = \{1, \ldots, k_j\}$
Parameters
m = number of potential investment projects
h = planning periods
p_w^τ = equals 1, if project w is realized in τ period, 0 otherwise
r_{jw} = equals 1, if project w is parallel department on j processing stage, otherwise 0
n = production steps
q = number of products
u_{jk}^τ = capacity of k parallel department in j processing stage in τ planning period
o_{ij} = equals 1, if product i is output of j processing stage
d_i^τ = cumulative demand for product i in τ period
s_{ij} = consumption per unit of the i product the j department
Decision variable
x_{ijk}^τ = quantity of the gross output of intermediate product i produced by k parallel department in j stage in τ period
g_{ij}^τ = variable determining size of sale of the product i produced by j processing stage in τ year

Resources and technical dependency This part consists of two groups of equations. First includes balances of the enterprise manufacturing capacities and flow of materials. It allows to determine size of the total production and sale achieved by enterprise. The second group of equations are financial equations.

Balance of capacities and flow of materials are stored in the form of the following constraints:

1. capacity constraints

- for any period τ the cumulative demand for capacity in department k not greater then available capacity in this department

$$\sum_{i \in I} x_{ijk}^\tau \leq u_{jk}^\tau; j \in J, k \in K_j, \tau \in T \qquad (7)$$

- for any period τ the cumulative demand for capacity in investment w not greater then available capacity for this investment

$$\sum_{i \in I} x_{ijw}^\tau r_{jw} \leq u_{jw}^\tau p_w^\tau; j \in J, w \in W, \tau \in T \qquad (8)$$

- for each product the cumulative sale of product i should not exceed demand.

$$\sum_{j \in J} o_{ij} g_{ij}^{\tau} \le d_i^{\tau}; i \in I \tag{9}$$

2. precedence constraints

- for each pair of consecutive production steps $(j, next(j))$ output j should equal input of $next(j)$.

$$\sum_{i \in I} \sum_{k \in K_j} x_{ijk}^{\tau} + \sum_{w \in W} \sum_{i \in I} x_{ijw}^{\tau} r_{jw} = \sum_{i \in I} \sum_{k \in K_{next(j)}} x_{i,next(j),k}^{\tau} c_{i,j} +$$
$$\sum_{w \in W} \sum_{i \in I} x_{i,next(j),w}^{\tau} r_{next(j),w} c_{i,j} + \sum_{i \in I} o_{i,next(j)} g_{i,next(j)}^{\tau}; j = 1, \ldots, n-1 \tag{10}$$

3. variable non-negativity and integrality conditions

$$x_{ijk_w}^{\tau} \ge 0; i \in I, j \in J, k_w \in W_j, \tau \in T \tag{11}$$

$$x_{ijk}^{\tau} \ge 0; i \in I, j \in J, k_w \in W_j, \tau \in T \tag{12}$$

$$g_{ij}^{\tau} \ge 0; i \in I, j \in J, \tau \in T \tag{13}$$

The second set of equations of the model are financial equations. They are linear equations, which for all the above-mentioned parameters determined by equations (7)–(13), determine specific items of the company's balance sheet, P&L account and cash flows (NCF) used to calculate the NPV. As an example, an equation for calculating a company's gross profit is presented below.

$$ZB^{\tau} = P^{\tau} - C_1^{\tau} - C_2^{\tau} - C_3^{\tau}; \tau \in T \tag{14}$$

where

$$P^{\tau} = \sum_{i \in I} \sum_{j \in J} c_i^{\tau} g_{ij}^{\tau} \tag{15}$$

$$C_1^{\tau} = \sum_{j \in J} \sum_{i \in I} \sum_{k \in K_j} kz_{ij}^{\tau} X_{ijk}^{\tau} \tag{16}$$

$$C_2^{\tau} = r_k KRK^{\tau} + r_d KRD^{\tau} \tag{17}$$

$$C_3^{\tau} = \sum_{w \in W} \chi^{\tau}(w) - \xi^{\tau}(w) \tag{18}$$

The notation used in financial equations is presented in Table 3

Table 3 Notation: Financial equations

Parameters
c_i^τ—selling price for product i in period t
kz_{ij}^τ—variable cost of processing the product i by department j in τ period
r_d—long-term interest rate
r_k—short-term interest rate
Decision variable
ZB^τ = variable determining gross profit in period τ
KRK^τ = variable determining the value of short-term credit in year τ
KRD^τ = variable determining the value of long-term credit in year τ

The remaining financial equations express commonly known dependencies. A detailed presentation on them would considerably increase the size of the article. Therefore, they are omitted.

Uncertainty of model parameters In the next step, the appropriate model of uncertainty is assigned for every parameters. The approach presented in this paper assumes that there are two ways of description of uncertainty—probabilistic and possibilistic. By using random variables, an assumption that the historical data of a given parameter are able to reflect the future behavior of this parameter should be satisfied. However, this assumption cannot always be met. Sometimes, there is no appropriate historical data or there are some human factors such us taxes, export quotas, government regulation which contribute to uncertainty of parameters. Thus, in some situations, the investors may not think that the past data of the parameter can well reflect its future behavior. They may like to use experts knowledge and their own experience to evaluate the future behavior of the parameter. In this situation fuzzy variable can be employed.

In the proposed model, material consumption and product cost are characterized by fuzzy numbers. Demand and selling prices are described by probability distribution. Then, the proposed framework employs hybrid simulation which allow different representations of uncertainty (e.g., by probability distributions, fuzzy numbers) to be processed according to their nature.

Dependency layer In capital budgeting, two kind of dependency exists—statistical and economical. Our framework takes into account economic dependencies in the process of selection of an effective portfolio of investment projects. Let us present this model using an example of steel industry. Let us assume, for example, that the modernization of an existing cold rolled sheet mill or the building of a new instead of old one will improve quality and extend range of cold-rolled steel sheets. Thus, the quality and range of hot-rolled steel sheets which are the product of following process steps will also be better. So, the sales of both products may increase. So the effects must be taking into consideration in PEM model.

Statistical dependency is used for describing relation between model parameters. Economic problems often involve parameters that are mutually correlated. For example, there is a correlation between enterprise product prices and raw material prices or between volumes of sales of different assortments. The omission of these dependencies leads to systematic errors in efficiency quantification, usually large overestimation of the actual efficiency. Mathematical model of statistical dependency is connected with modeling of uncertainty of model parameters. Dependency between parameters characterized by fuzzy numbers are described by interval regression. Interval regression is an extension of the classical (crisp) regression where regression parameters are bounded closed intervals. For probabilistic parameters their dependency is determined by the correlation matrix. To process them, a method presented by Yang [37] based on Cholesky decomposition of the correlation matrix is utilized.

4.4 Procedure of Determining Portfolio Evaluation Model

The proposed procedure of determining the effectiveness of investment portfolio consists of two stages. It combines the procedure of stochastic simulation with execution of arithmetic operations on interactive fuzzy numbers. To execute such arithmetic operations non-linear programming is used. Computation procedure in this case is the following. Random variable values are drawn from among mentioned above parameters expressed in the form of the probability distribution. The procedure of generation accounts statistical dependency between variables. These values and remaining parameters expressed in the form of fuzzy numbers allow to determine evaluation parameter as fuzzy number. The problem of determining the fuzzy number characterizing evaluation parameter may be written owing to use of the concept of α-levels of fuzzy sets. Thus, the variables y corresponding to the parameters that are expressed in the form of fuzzy numbers are introduced, and then the parameters are replaced for those variables. Additionally, the following constraints are imposed:

$$\inf\left(\tilde{Y}_i\right)_\alpha \le y_i \le \sup(\tilde{Y}_i)_\alpha \tag{19}$$

$$y_i \ge \inf\left(a_1^{iz}\right) \cdot y_z + \inf\left(a_2^{iz}\right) \tag{20}$$

$$y_i \le \sup\left(a_1^{iz}\right) \cdot y_z + \sup\left(a_2^{iz}\right) \tag{21}$$

where

- $\inf\left(\tilde{Y}_i\right)_\alpha$, $\sup(\tilde{Y}_i)_\alpha$—respectively lower and upper bounds of α-level of the fuzzy parameter $\tilde{Y}_i$
- $\sup\left(a_1^{iz}\right)$, $\inf\left(a_1^{iz}\right)$, $\sup\left(a_2^{iz}\right)$, $\inf\left(a_2^{iz}\right)$ —respectively lower and upper bounds of interval regression coefficients describing dependency between parameters $\tilde{Y}_z$ and $\tilde{Y}_i$

Algorithm 1 Procedure of determining evaluation model

$n \leftarrow 1$;

2: Randomly generate vector probabilistic variables taking into account the correlation between them

$\alpha = 0$;

4: Define α-levels for fuzzy variables defining efficiency parameter

Calculate (sup) and (inf) for defined α-levels by finding: $eff_\alpha \leftarrow min$ and $eff_\alpha \leftarrow max$ under the problem constrains specified by constraints

6: $\alpha = \alpha + \phi$

If $\alpha \leqslant 1$ goto **STEP 4** else $n = n + 1$

8: if $n \leqslant n_{max}$ goto **STEP 2**

Calculate mean value, standard deviation, and lower and upper cumulative distributions of the NPV.

Next, in order to determine the lower and upper bound of the respective α-level of the efficiency parameter, the following constrained optimization problems must be solved:

$$NPV_\alpha \longrightarrow min \qquad (22)$$

for the definition of the lower bound of the α-level of the NPV,

$$NPV_\alpha \longrightarrow max \qquad (23)$$

for the definition of the upper bound of the α-level of the NPV. Drawing probabilistic values and determining NPV is repeated n_{max} times. As result n_{max} fuzzy sets characterized by membership functions $(\mu_1^{NPV}, \dots, \mu^{NPV})$ are obtained. In this case NPV is represented by a random fuzzy variable. Based on the vector $(\mu_1^{NPV}, \dots, \mu^{NPV})$, the mean value, standard deviation as well as lower and upper cumulative distributions for the NPV are calculated. The hybrid procedure which implements the described approach is presented in the following algorithm.

5 Numerical Example

The capital budget was determined for the production process presented on the Fig. 1. This setup includes the production cycle in steel industry, from production of the pig iron, production of steel, hot rolling products to production products coated with metal and plastics.

Five investment projects are taken into consideration: steel making plant, hot rolled sheet mill, cold-rolled sheet mill, hot-dip galvanizing sheet plant and sheet organic coating plant. In Fig. 1 they are denoted with the suffix -project. Parameters for the estimation of efficiency and the risk of investment projects in case of the investment in iron metallurgy are: quantity and selling prices, costs of materials and quantity of investment outlays. It was recognized also from here, that in the

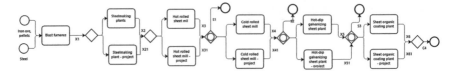

Fig. 1 Diagram of the analyzed technological setup

simulation experiment is necessary to take into consideration the uncertainty of the possible quantity of sale for each of products ranges being produced by the company, prices of these products, prices of metallurgic raw materials (prices of iron ores and the pellets), consumption per unit indexes, quantity of investment outlays. It was assumed, that remaining parameters of the efficiency calculus were determined. Prices of individual assortments of metallurgic products and metallurgic raw materials are correlated strongly. Similarly, sale quantities of each assortment of metallurgic products are correlated. This fact was taken into consideration when processing the values of efficiency calculus uncertain parameters.

In the computational experiment it was taken into consideration the uncertainty of the possible quantity of sales for each of products ranges being produced by the company, prices of these products, prices of semi-finished steel (prices of continuous casting stands), investment outlay for projects, construction period for investment projects and consumption per unit indexes for all products. Sales and possible quantity of sale for each product are described be probability distribution (in this case normal distribution). Rest of parameters were presented as a triangular fuzzy numbers.

In numerical example, we identify following dependencies:

- Prices of individual assortments of metallurgic products and metallurgic raw materials are correlated. Also sale quantities of each assortment of metallurgic products are correlated. Those parameters are using to describe benefit interdependency.
- Equations of manufacturing capacity balance are using for describing resource dependency and technical dependency.

Material consumption as well as product and half-product prices are given in the form of fuzzy numbers. They are presented, respectively, in Tables 4 and 5.

Sale parameters are given by normal probability distributions given in Table 6.

Table 4 Trapezoidal fuzzy numbers (TFN) indicating material consumption

Material consumption	TFN
Steel half-products—molten iron	(0.855, 0.860, 0.870, 0.875)
Half-products—hot rolled steel sheets	(1.058, 1.064, 1.075, 1.078)
Hot rolled steel sheets—cold rolled sheets	(1.105, 1.111, 1.124, 1.130)
Cold rolled sheets—dip galvanized sheets	(1.010, 1.020, 1.026, 1.031)
Dip galvanized sheets—organic coated sheets	(0.998, 0.999, 1.000, 1.001)

Table 5 Trapezoidal fuzzy numbers (TFN) for prices

Price	TFN (USD/t)
Iron ore	(335, 360, 400, 425)
Lumps	(375, 400, 440, 470)
Steel scrap	(940, 960, 1010, 1035)
Hot rolled sheets	(2040, 2080.8, 2177.7, 2228.7)
Cold rolled sheets	(2220.075, 2266.65, 2370.15, 2427.075)
Hot dip galvanized sheets and strips	(2535.75, 2588.25, 2709, 2772)
Organic coated sheets and tapes	(3450.6, 3519.825, 3684.9, 3754.125)

Table 6 Probability distributions indicating sale parameters

Sale	Mean value	Std. dev.
Hot rolled sheets	4704.0	117.5
Cold rolled sheets	2750.0	51.4
Hot dip galvanized sheets and tapes	1147.9	52.4
Organic coated sheets and tapes	708.4	30.8

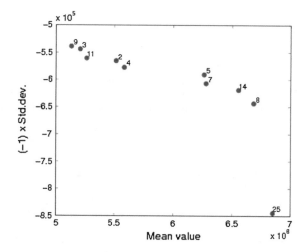

Fig. 2 Pareto optimal solutions for the problem of selection of an efficient portfolio of investment projects

The Cholesky matrix which describes the dependencies between sale parameters is given by the Eq. 24.

$$\begin{pmatrix} 1.00000 & 0.87786 & 0.91142 & 0.86321 \\ 0.00000 & 0.47891 & 0.24007 & 0.27276 \\ 0.00000 & 0.00000 & 0.33418 & 0.34165 \\ 0.00000 & 0.00000 & 0.00000 & 0.25249 \end{pmatrix} \qquad (24)$$

For the computational example, the α-level for fuzzy variables are set at 10 and the number of simulation are set to 100. The result for the computational example is shown in Fig. 2

6 Conclusion

This paper presents a new framework for selecting an effective portfolio of investment project. The presented mathematical model and developed algorithm allow to generate a set of Pareto optimal solutions. The solutions on Pareto front represent different variations of a company's acceptable capital budgets along with the estimation of their effectiveness (expected value of financial evaluation parameter) and risk (standard deviation of financial evaluation parameter). The main advantages of the proposed framework over existing ones is that it allows to take into account statistical as well as economic dependencies between projects. It also allows for flexible definition of uncertainty of the parameters using probability distributions or fuzzy numbers.

References

1. Basiura, B., Duda, J., Gawe, B., Opia, J., Peech-Pilichowski, T., Rbiasz, B., Skalna: Advances in Fuzzy Decision Making: Theory and Practice, Springer, (333), (2015)
2. Rebiasz, B., Gawel, B., Skalna, I.: Hybrid framework for investment project portfolio selection. In Computer Science and Information Systems (FedCSIS), 2014 Federated Conference, 1117–1122 (2014).
3. Baudrit, C., Dubois, D, Guyonet, D., (2006), Joint Propagation and Exploitation of Probabilistic and Possibilistic information in Risk Assessment. IEEE Transaction on Fuzzy Systems. **14**, 593–607 (2006)
4. Lorie J.H, Savage L.J.: Three problems in capital rationing. Journal of Business. **28**, 229–239 (1955)
5. Guyonnet, D., Bourgine, B., Dubois, D., Fargier, H., Cme, B., Chils P.J.: Hybrid Approach for addressing uncertainty in risk assessment. Journal of Environmental Engineering. **126**, 68–76 (2003)
6. Cooper, A., Ferson, S., Ginzburg, L.: Hybrid processing of stochastic and subjective uncertainty data. Risk Analysis. **16**, 785–791 (1996)
7. Hasuike, T., Katagirib, H., Ishii, H.: Portfolio selection problems with random fuzzy variable returns. Fuzzy Sets and Systems. **160**, 2579–2596 (2009)
8. Couso, I., Dubois, D., Montes, S., Sanchez, L.: On various definitions of the variance of a fuzzy random variable, In: Proceedings of 5th International Symposium on Imprecise Probabilities and Their Applications, Prague, Czech Republic (2007)
9. Liang, R., Gao, J.: Dependent-Chance Programming Models for Capital Budgeting in Fuzzy Environments. Tsinghua Science and Technology. **13**:1, 117–120 (2008)
10. Feng, Y., Hu, L., Shu, H.: The variance and covariance off uzzy random variables and their applications. Fuzzy Sets and Systems. **120**, 487–497 (2001)
11. Rębiasz, B.: Selection of efficient portfolios-probabilistic and fuzzy approach, comparative study. Computers & Industrial Engineering. **64**:4, 1019–1032 (2013)

12. Ferson, S., Ginzburg, L.R.: Difference method are needed to propagate ignorance and variability. Reliability Engineering System Safety. **54**, 133–144 (1996)
13. Bradley, S.P., Frey, C.: Equivalent Mathematical Programming Models of Pure Capital Rationing, Journal of Financial and Quantitative Analysis. **6**, 345–361 (1978)
14. Carleton, W.T.: Linear programming and Capital Budgeting Models: A New Interpretation. Journal of Finance. **23**, 825–833 (1974)
15. Ignazio, J.P.: An approach to the Capital Budgeting Problem with Multiple Objectives. The Engineering Economist. **21**, 259–272 (1976)
16. Archer, N.P., Ghasemzadeh, F.: An integrated framework for project portfolio selection. International Journal of Project Management. **17**:4, 207–216 (1999)
17. Markowitz, H.M.: Portfolio Selection Efficient Diversification of Investment. Wiley, New York, 1959
18. Seitz, N.E.: Capital Budgeting and Long-Tern Financing Decisions, 3rd ed. Dryden Press, USA (1999)
19. Acharaya, P.K.De.D., Sahu K.C.: A Chance-Constrained Goal Programming Model for Capital Budgeting. The Journal for the Operational Research Society. **33**:7, 635–638 (1982)
20. April, J., Glover, F., Kelly, J.P.: OPTFOLIO - A Simulation Optimization System For Project Portfolio Planning. In: Proceedings of the 2003 Winter Simulation Conference, pp. 301–309 (2003)
21. Badri, M.A., Davis, D., Davis, D.: A comprehensive 0-1 goal programming model for project selection. International Journal of Project Management. **19**, 243–252 (2001)
22. Olson, D.L.: Decision Aids for Selection Problems. Springer Series in Operations Research, New York (1996)
23. Huang, X.: Credibility-based chance-constrained integer programming models with fuzzy parameters. Information Science. **176**:18, 2698–2712 (2006)
24. Huang, X.: Fuzzy chance-constrained portfolio selection. Applied Mathematics and Computation. **177**:2, 500–507 (2006)
25. Liu, B., Iwamura, K.: Chance constrained programming with fuzzy parameters. Fuzzy Sets and Systems. **94**:2, 227–237 (1998)
26. Kahraman, C., Ruan, D., Dozdag C.E.: Optimization of Multilevel Investments Using Dynamic Programming Based on Fuzzy Cash Flows. Fuzzy Optimization and Decision Making. **2**:2; 101–122 (2003)
27. Dickinson, M.W., Thomton, A.C., Graves, S.: Technology portfolio management. Optimizing interdependent projects over multiple time period. IEE Transaction on Engineering Management. **48**:4, 518–527 (2001)
28. Santhanam, R., Kyparisis, G.J.: A decision model for interdependent information system project selection. European Journal of Operational Research. **89**:2, 380–399 (1996)
29. Zuluaga, A., Sefair J., Medaglia A.: Model for the Selection and Scheduling of Interdependent Projects. Proceedings of the 2007 Systems and Information Engineering Design Symposium, University of Virginia., 2007. http://wwwprof.uniandes.edu.co/~amedagli/ftp/PMAfternoonSession1T5-04.pdf).
30. Medaglia, A.L., Graves, S.B., Ringuest, J.L.: A multiobjective evolutionary approach for linearly constrained project selection under uncertainty. European Journal of Operational Research. **179**:3, 869–894 (2007)
31. Hall, D.L., Nauda, A.: An interactive approach for selecting IR&D projects. IEEE Trans. Eng. Management. **37**:2, 126–133 (1990)
32. Aaker, D.A., Tyebjee, T.T.: A model for the selection of interdependent R&D projects. IEEE Transactions on Engineering Management. **25**:2, 30–36 (1978)
33. Fox, G.E, Baker, N.R., Bryant, J.L.: Economic models for R and D project selection in the presence of project interactions. Management science. **30**:7, 890–902 (1984)
34. Eilat, H., Golany, B., Shtub, A.: Constructing and evaluating balanced portfolios of R&D projects with interactions: A DEA based methodology. European Journal of Operational Research. **172**:3, 1018–1039 (2006)

35. Carlsson, C., Fullér, R.: Fuzzy multiple criteria decision making: Recent developments. Fuzzy sets and systems. **78**:2, 139–153 (1996)
36. Rębiasz, B., Gaweł, B., Skalna, I: Fuzzy multi-attribute evaluation of investments. In: Proceedings of IEEE Federated Conference on Computer Science and Information Systems (FedCSIS), (2013)
37. Yang, I.-T.: Simulation-based estimation for correlated cost elements. International Journal of Project Management. **23**:4, 275–282 (2005)
38. Bernhard, R.H.: Mathematical programming models for capital budgeting-survey, generalization and critique. Journal of Financial and Quantitative Analysis. **4**, 111–158 (1969)
39. Lusztig P, Schwab B. A Note of the Application of Linear Programming to Capital Budgeting. Journal of Financial and Quantitative Analysis. **3**, 427–431 (1968)

Towards Predicting Stock Price Moves with Aid of Sentiment Analysis of Twitter Social Network Data and Big Data Processing Environment

Andrzej Romanowski and Michał Skuza

Abstract This chapter illustrates design and evaluation of a sentiment analysis based system that may be used to predict future stock prices. Social media information is processed in order to extract opinions that are associated with Apple Inc. company. The authors took advantage of large datasets available from Twitter micro blogging platform and widely available stock market records. Data was collected during 3 months and processed for further analysis. Machine learning was employed to conduct sentiment classification of data in order to estimate future stock prices. Calculations were performed in distributed environment according to Map Reduce programming model. Evaluation and discussion of predictions results for different time intervals and input datasets is discussed in terms of efficiency and feasibility of the chosen approach.

Keywords Sentiment analysis · Big data processing · Social networks analysis · Stock market prediction

1 Introduction

Recent years have shown not only an explosion of data, but also widespread attempts to process, analyse and interpret it for various practical reasons. Computer systems operate on data measured in terabytes or even petabytes and both users and computer systems constantly generate the data at incredible pace. Scientists and computer engineers have created special term "big data" to name this trend. Main features of big data are volume, variety and velocity (some add veracity, as well). Volume stands for large sizes, which cannot be easily processed with traditional database systems and single machines. Velocity means that data is constantly created at a fast rate and variety corresponds to different forms such as text, images and videos. One of

A. Romanowski (✉) · M. Skuza
Lodz University of Technology, Institute of Applied Computer Science,
Stefanowskiego 18/22 str, 90-924 Lodz, Poland
e-mail: androm@kis.p.lodz.pl

© Springer International Publishing AG 2017
T. Pełech-Pilichowski et al. (eds.), *Advances in Business ICT: New Ideas from Ongoing Research*, Studies in Computational Intelligence 658,
DOI 10.1007/978-3-319-47208-9_7

possible, interesting applications illustrated in this work is the attempt to use big data for financial analysis, i.e. stock price prediction. The analyses of information that are by-products of different business activities by companies can lead to better understanding the needs of their customers and prediction future trends. It was previously reported in several research papers that precise analysis of trends could be used to predict financial markets [1, 2]. Although financial markets researchers are not unanimous on possibility of stock prediction in general, but most of them agree to the efficient-market hypothesis suggesting that stock prices reflect all currently available information [1].

As the social media networks get more popular and gain more and more attention from variety of stakeholders, including individuals, officials, organisations and public opinion in general, the data or the information available in these services reflect the real life and economy more and more as well. Therefore, it is natural to try to use this data and information in order to infer about possible relationship between markets and opinions shared in the network and moreover, the impact that this information could have on the market. We chose the Twitter micro blogging service as the source of data and hadoop map reduce based platform for processing it. The last element is the way information could be processed for extraction of useful clues that may reflect stock behaviour. We chose sentiment analysis of twitter messages associated with Apple company to provide a proof-of-concept study of our approach. The details and reasons behind them are given in later sections of this chapter.

1.1   Big Data

There are several definitions what "Big Data" is, one of them is following: "Big data refers to datasets whose size is beyond the ability of typical database software tools to capture, store, manage, and analyse" [3]. This definition emphasizes key aspects of big data that are volume, velocity and variety [4]. According to IBM reports [5] everyday "2.5 quintillion bytes of data" is created and that figure is constantly increasing. This is due to previously described ubiquitous access to the Internet and growing number of devices. Data is created and delivered from various systems operating in real-time. For example social media platforms aggregate constantly information about user activities and interactions e.g. one of most popular social sites Facebook has over 1 billion daily active users as for Q1 2016 that rises nearly linearly (with more than 600 millions at the end of 2012). Output rate of the system can be also important when nearly real-time analyses are needed in cases such for recommendations systems when the user's input affects content provided by web site. This aspect requires variety of manners in order to efficiently and flexibly store the data to maximize response speed. Big data is rarely well structured hence column-oriented database schemes or one of schema-less systems (NoSQL) are becoming common approaches.

There are many reasons for such rise of big data. One of them is the increasing number of mobile devices such as smartphones, tablets and computer laptops all connected to the Internet. It allows millions of people to use web applications and services that create massive amounts of logs of activity, which in turn are gathered and processed by companies. Another reason is that computer systems started to be used in many sectors of the economy from governments and local authorities to health care to financial sector. Big size of data and the fact it is generally not well structured result in situation that conventional database systems and analysis tools are not efficient enough to handle it. In order to tackle this problem several new techniques ranging from in-memory databases to new computing paradigms were created. Besides big size, the analysis and interpretation are of main concern and application for big data perspective stakeholders. Analysis of data, also known as data mining, can be performed with different techniques such as machine learning, artificial intelligence and statistics. And again it is important to take into consideration the size of data to be processed that in turn determines if a given existing algorithm or approach is applicable. However, big data is not only challenging but creates enormous opportunities as well and our example given in this chapter regards the possible extraction of stock price trends from it.

1.2   Social Media

One of the trends leading to rise of big data is Web 2.0, i.e. a shift from static websites to interactive, user-generated content (UGC) based ones. UGC manifested in many services such as blogging, podcasting, social networking and bookmarking. Users can create and share information within open or closed communities and by that contribute to volumes of big data. Web 2.0 led to creation of social media that now are means of creating, contributing and exchanging information with others within communities by electronic media. Social media can be also summarized as "built on three key elements: content, communities and Web 2.0" [6]. Each of those elements is a key factor and is necessary for social media. One of the most important factors boosting social media is increasing number of always Internet-connected mobile devices such as smartphones and tablets.

Twitter is a micro blogging platform, which combines features of blogs and social networks services. Twitter was established in 2006 and experienced rapid growth of users in the first years of operations. Currently it has over 640 million registered users and over 300 million active monthly users [7] as for begging of 2016. Registered users can post and read messages called "tweets"; each up to 140 Unicode characters long –originated from SMS carrier limit. Unregistered users can only view tweets. Users can establish only follow or be-followed relationships. A person who subscribes to other user is referred as "follower" and receives real-time updates from that person. However users do not have to add people who are their followers. Twitter can be accessed from various services such as official Twitter web page, mobile applications from third parties and SMS service. As Twitter is an extremely widespread service, especially in US and as the data structure is compact so it forces users to post short comments authors of this paper believe this is a good source of information in the sense of snapshots of moods and feelings as well as for up-to-date events and current situation commenting. Moreover, Twitter is a common PR communication tool for politicians and other VIPs shaping, or having impact on the culture and society of large communities of people. This is the case in most of the regions, but especially US is a country that boosted twitter micro blogging since the SMS (short text messaging) system was not that popular beforehand. Therefore Twitter was chosen for experimental data source for this work on predicting stock market.

2   Predicting Future Stock Prices

Main goal of this section is to describe implementation of a proposed concept for a system predicting future stock prices based on opinion detection of messages from Twitter micro blogging platform.

2.1   Stock Market Company Selection

Unlike the authors of [8] we chose Apple Inc. – a well known consumer electronics company – a leader the world's biggest companies list according to Forbes magazine as of 2016 (and for several consecutive years now) by the both market values (way ahead of the second on the list) and brand value. The reasons behind this decision are two fold. First of all the variety of consumer goods produced by Apple and therefore being one of the most known labels makes it more or less know recognizable to many possible Twitter bloggers. The enormous value and profits of the company make it a common subject of discussion among both ordinary people as well as various financial and other VIP authorities and hence we counted to have more quality data to be analysed for Apple than for any other company.

2.2   Stock Data Retrieving

The most important piece of information in order to both process and compare the stock values it to get the ground truth, i.e. the actual results of the stock market. Intraday stocks market data was retrieved from Yahoo Finance in CSV for-mat. Stock quotes are collected on a per minute basis. NASDAQ stock market allows trading stocks only from 9:30 a.m. to 4:00 p.m. in Eastern Standard Time (EST).

Intraday stocks data are saved to database with corresponding stock sym-bols. Retrieval time is converted into Central European Summer Time (CEST). Snapshot of Apple Inc. (AAPL) stock chart and its equivalent data (that are also available in easily convertible CSV format file) is presented at Fig. 1.

Fig. 1 Snapshot of Apple Inc. (AAPL) stock chart taken from Nasdaq

2.3 Experimental System Design and Implementation

System design is presented on Fig. 2 and it consists of four main components: Retrieving Twitter data, pre-processing and saving to database (1), stock data retrieval (2), model building (3) and predicting future stock prices (4). Each component is described later in this text.

1. Retrieving Twitter data, pre-processing and saving to database.
   This component is responsible for retrieving, pre-processing data and preparing training set. There are two labelling methods used for building training set: manual and automatic.
2. Stock data retrieval.
   Stock data is gathered on a per minute basis. Afterwards it is used for estimating future prices. Estimation is based on classification of tweets (using sentiment analysis) and comparing with actual value by using Mean Squared Error (MSE) measure.

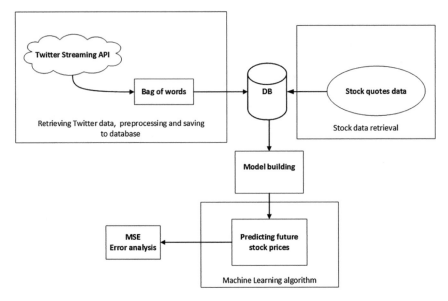

Fig. 2 Design of the system

3. Model building.
   This component is responsible for training a binary classifiers used for sentiment detection.
4. Predicting future stock prices.
   This component combines results of sentiment detection of tweets with past intra-day stock data to estimate future stock values.

2.4 Twitter Data Acquisition and Pre-processing

Twitter messages are retrieved in real time using Twitter Streaming API. Streaming API allows retrieving tweets in quasi-real time (server delays have to be taken into consideration). There are no strict rate limit restrictions, however only a portion of requested tweets is delivered. Streaming API requires a persistent HTTP connection and authentication. While the connection is kept alive, messages are posted to the client. Streaming API offers possibility of filtering tweets according to several categories such as location, language, hashtags or words in tweets. One disadvantage of using Streaming API is that it is impossible to retrieve tweets from the past this way.

Tweets were collected over 3 months period from 2nd January 2013 to 31st March 2013. It was specified in the query that tweets have to contain name of the company or hashtag of that name. For example in case of tweets about Facebook Inc. following words were used in query 'Apple', '#Apple', 'AAPL' (stock symbol of the company) and '#AAPL'. Tweets were retrieved mostly for Apple Inc. (traded as 'AAPL') in order to ensure that datasets would be sufficiently large for classifications. Retrieved data contains large amounts of noise and it is not directly suitable for building classification model and then for sentiment detection. In order to clean twitter messages a program in Python programming language was written. During processing data procedure following steps were taken. Language detection information about language of the tweet is not always correct. Only tweets in English are used in this research work. Duplicate removal - Twitter allows to repost messages. Reposted messages are called retweets. From 15 to 35 % of posts in datasets were retweets. Reposted messages are redundant for classification and were deleted. After pre-processing each message was saved as bag of words model – a standard technique of simplified information representation used in information retrieval.

2.5 Sentiment Analysis

Unlike classical methods for forecasting macroeconomic quantities [1, 8–10] prediction of future stock prices is performed here by combining results of sentiment classification of tweets and stock prices from a past interval [11]. Sentiment analysis [12, 13] - also known as opinion mining refers to a process of extracting information about subjectivity from a textual input. Sentiment analysis usually refer to natural language processing in terms of text analysis and interpretation and computational linguistics in order to extract, identify, characterize the sentiment (an associated point of view, or attitude towards sth.) of a given text. In other words the goal of the process is to determine if whether a given text snippet is negative, positive or neutral. In order to achieve this it combines techniques from natural language processing and textual analysis. Capabilities of sentiment mining allow determining whether given textual input is objective or subjective. Polarity mining is a part of sentiment in which input is classified either as positive or negative.

In order to perform a sentiment analysis classification a model has to be constructed by providing training and test datasets. One way of preparing these datasets is to perform automatic sentiment detection of messages. This approach was used in several works such as [14]. Another possibility of creating training and test data is to manually determine sentiment of messages, which means it is a standard, supervised learning approach. Taking into consideration large volumes of data to be classified and the fact they are textual, Naïve Bayes method was chosen due to its fast training process even with large volumes of training data and the fact that is it is incremental. Considered large volumes of data resulted also in decision to apply a map reduce version of Naïve Bayes algorithm. In order to perform sentiment analysis on prepared bags of words a model has to be constructed by providing training and test datasets for classification. These datasets were created using two different methods. One was applying an automatic sentiment detection of messages. It was achieved by employing SentiWordNet [15] which is a publicly available resource aimed to support performing sentiment and opinion classifications. The other method was a manual labelling of sentiment of tweets. Each message was marked as positive, negative or neutral. There were two training datasets. First one consisted of containing of 800 hundred tweets. The other dataset consisted of 2.5 million messages. Only 90 % of each dataset was used directly as a training set the other 10 % was used for testing.

As a result of two classifiers were obtained using manually labelled dataset. First classifier determines subjectivity of tweets. Then polarity classifier classifies subjective tweets, i.e. using only positive and negative and omitting neutral ones. In order to use classification result for stock prediction term: 'sentiment value' (denoted as a ε) was introduced - it is a logarithm at base 10 of a ratio of positive to negative tweets (Eq. 1).

$$\varepsilon : \log_{10} \frac{number\_of\_positive\_tweets}{number\_of\_negative\_tweets} \tag{1}$$

If ε is positive then it is expected that a stock price is going to rise. In case of negative ε it indicates probable price drop. In order to estimate price of stock, classification results are combined with a linear regression of past prices where one weight is a sentiment value. Predictions for a specific time point are based on analysis of tweets and stock prices from a past interval - Eq. 2 shows the formula for the relationships of past value of stock taken into analysis.

$$y_i = \alpha + (\beta + \varepsilon_i)x_i \tag{2}$$

where: y_i is a past value of a stock at a given time of x_i, x_i is time variable, ε_i is a sentiment value calculated for a given time of x_i, $i = 1, ..., n$, β is a linear regression coefficient defined as (Eq. 3)

$$\beta = \frac{\Sigma x_i y_i - \frac{1}{n} \Sigma x_i \Sigma y_i}{\Sigma x_i^2 - \left(\frac{1}{n} \Sigma x_i\right)^2} \tag{3}$$

and α coefficient is given by (Eq. 4):

$$\alpha = \bar{p} - \beta \bar{t} \tag{4}$$

where $\bar{p}$ and $\bar{t}$ are mean values of price of stock over a period of t.

3   Results and Analysis

Predictions were prepared using two datasets for several different time intervals, i.e. time differences between the moment of preparing the prediction and the time point for which the forecast was prepared. Predictions were conducted for 1 h (i.e. 60 min), half an hour (30 min), 15 and 5 min ahead of the moment being forecasted.

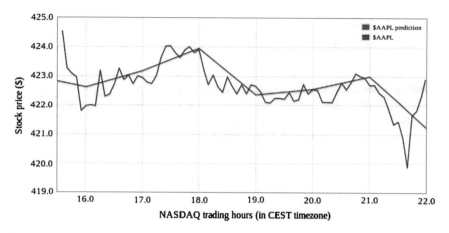

Fig. 3 One-hour prediction. Manually labelled 'AAPL' training dataset

Two tweet datasets were used: one with messages containing company stock symbol 'AAPL', and the other dataset included only tweets containing name of the company, i.e. 'Apple'. Training datasets consisted of 3 million tweets with stock symbols and 15 million tweets with company name accordingly. Tweets used for predictions were retrieved from 2nd to 12th of April 2013. Approximately 300,000 tweets were downloaded during New York Stock Exchange trading hours each day via Twitter Streaming API.

Experiments were conducted using two models of classifiers, first was built using manually labelled, dataset-based trained classifier and the other was trained with automatically labelled tweet training datasets. Experiments were conducted in the following manner. For each of the following prediction time intervals: 1 h, 30 and 15 min all four models (permutations of manual or auto-classifiers coupled with AAPL or Apple keywords) were used. For 5 min prediction small number of tweets with stock symbol (AAPL) per time interval resulted in limiting predictions only to models trained with messages with company name (Apple).

Time axis shown on Fig. 3 and the following figures shows NASDAQ trading hours converted to CEST 1 time zone. Sample predictions are presented for 1 h, 30, 15 and 5 min, and the corresponding figures shows plots predicted and actual stock prices.

3.1 One-Hour Predictions

A first objective was to test a performance of predictions for 1-h intervals. Example results of 1-h interval predictions are presented on Fig. 3.

Blue line (fluctuated) corresponds to actual stock prices and red line (steadily changing) shows predicted values. Predictions in this time intervals would not provide accurate result but they can be used to evaluate if the method correctly estimates trends. Predictions for the same day are presented below using 4 different models of classifier.

As it can be observed on Figs. 3, 4, 5 and 6 predictions of each model perform similarly. They correctly forecast trends, i.e. reflect the general direction of the stock price development and significant changes in the stock price movements somehow averaged over time. However, due to a relatively substantial time interval, i.e. 1 h period advance, it is not possible to determine whether models can predict sudden (i.e. occurring quickly, unexpectedly or without any long term notice) price movements. Furthermore when comparing results for two datasets, predictions using models trained with tweets with stock symbol perform better that those trained with tweets containing full company name.

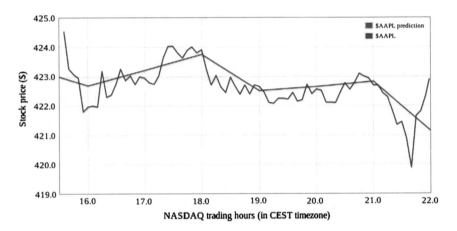

Fig. 4 One-hour prediction. Automatically labelled 'AAPL' training dataset

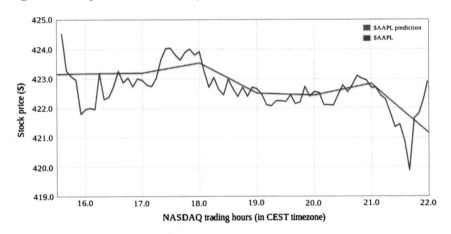

Fig. 5 One-hour prediction. Manually labelled 'Apple' training dataset

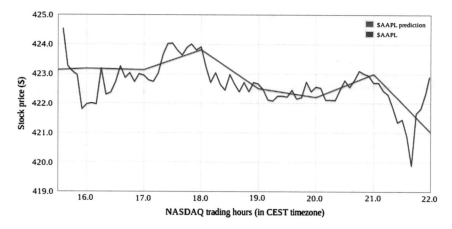

Fig. 6 One-hour prediction. Automatically labelled 'Apple' training dataset

3.2   30 min Predictions

This subsection describes 30 min predictions. It is anticipated for these predictions not only to forecast trends but additionally certain sudden price movements to be predictable as well. This expectation is due to the fact of smaller time interval between time of prediction and forecasted moment. Results are shown on Figs. 7, 8, 9 and 10 graphs.

First model ('manual' for AAPL keyword) predictions are less accurate in comparison with second model (auto/AAPL). Yet, significant difference between predicted and actual prices from 17 to 21 in both first models is still there.

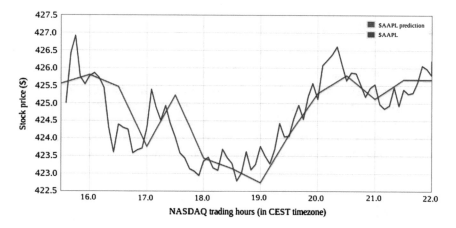

Fig. 7 30 min prediction. Manually labelled 'AAPL' training dataset

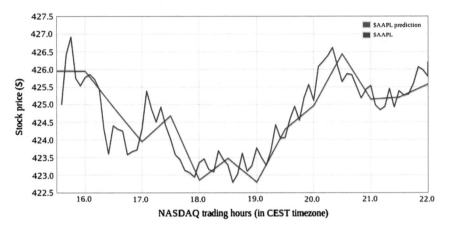

Fig. 8 30 min prediction. Automatically labelled 'AAPL' training dataset

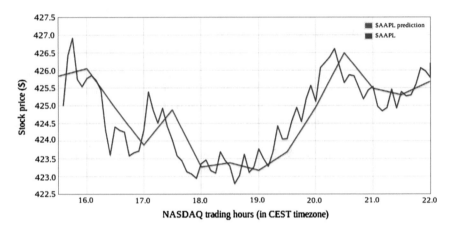

Fig. 9 30 min prediction. Manually labelled 'Apple' training dataset

Models using dataset with actual company name (plots at Figs. 9 and 10) perform much better than two first ones. Predictions of prices follow actual ones. However in all cases price forecasting is less accurate when there are several dynamic changes of price movement trend. It is especially visible in all figures for periods from 18 to 20 h that predictions do not show correlations with actual prices. It may result, among the others, from still too long time intervals in comparison to rapid price movements.

3.3   15 min Predictions

For 15 min dynamic price movements are somehow reflected in prediction, although it is not any significant indication in a sense of preserving real nature and amplitude

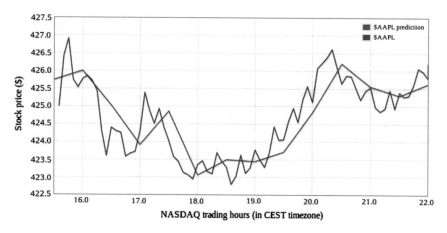

Fig. 10 30 min prediction. Automatically labelled 'Apple' training dataset

of those fluctuations. Rapid price movements are not easily indicated due to chosen time interval; still too long for better accuracy. Interestingly, for this 15 min interval classification based on tweets with stock symbol yield better results.

Furthermore it is important to note that using automatically trained training with bigger number of records strongly affects result of prediction.

There are somehow modeled rapid price movements in time interval from 19 to 21 in Figs. 11 and 12 whereas in Figs. 13 and 14 there are more smooth lines. Furthermore it is important to note that using automatically trained data set training with bigger number of records strongly affects result of prediction.

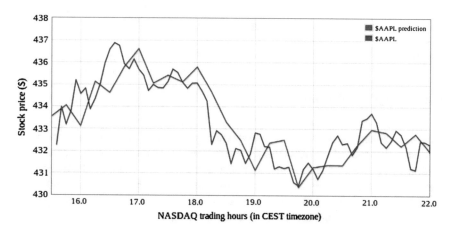

Fig. 11 15 min prediction. Manually labelled 'AAPL' training dataset

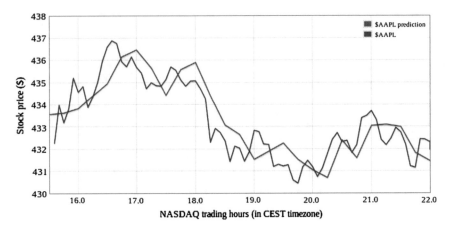

Fig. 12 15 min prediction. Manually labelled 'Apple' training dataset

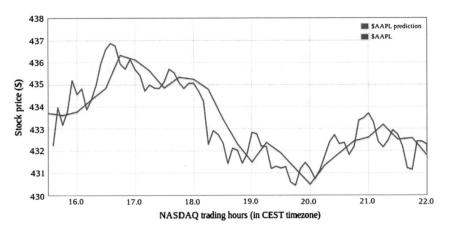

Fig. 13 15 min prediction. Automatically labelled 'AAPL' training dataset

3.4   5 min Predictions

Last experiment was to perform 5 min predictions. Due to short time interval it was expected that prediction would be the most accurate. In this part only dataset build with messages with actual company name was used. The reason behind dropping the AAPL tag data set is caused by the too small number of messages with stock symbol per time interval. So small number of messages would yield, in our opinion, unreliable results of predictions. Results are shown on Figs. 15 and 16.

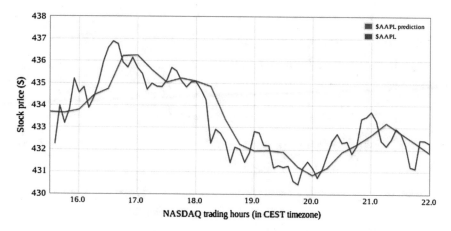

Fig. 14 15 min prediction. Automatically labelled 'Apple' training dataset

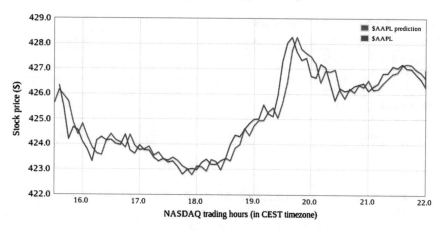

Fig. 15 5 min prediction. Manually labelled 'Apple' training dataset

For both Figs. 15 and 16 we can observe highly correlated results of predictions with a small delay in favour of actual results. One can say such a highly correlated plot for short advance period of 5 min can be regarded not only as an advantage proving performance of the proposed approach but at the same time can be treated as the drawback of the undertaken methodology, since it is an evidence linear regression coefficient playing role in Eqs. 1–4 for our model. It is difficult to authoritatively and unambiguously address this issue; partially we can agree with this point, but still both the theory underlying the stock predictions and promising practical opportunities advocate our approach to be conceivable choice in a range of applications.

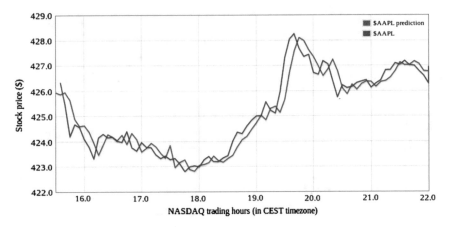

Fig. 16 5 min prediction. Automatically labelled 'Apple' training dataset

Table 1 Mean square error values of predicted and actual stock prices

MSE	1 h	30 min	15 min	5 min
Manual and 'AAPL'	1.5373	0.6325	0.3425	–
Auto and 'AAPL'	0.947	0.3698	0.2814	–
Manual and 'Apple'	1.9287	1.5152	0.9052	0.5764
Auto and 'Apple'	1.8475	1.4549	0.8325	0.3784

Finally the results are summarized for all four combinations of models and training methods in terms of error in comparison to the actual values. Results of predictions are compared to actual stock prices using Mean Square Error (MSE) measure that are presented in Table 1.

Perhaps along with the development of twitter social network, i.e. with increasing number of users, also the number of twits per time unit will rise and it will be feasible to analyse stock symbol based messages even for a short advance intervals. This may yield better results as shown for greater periods when AAPL based sentiment analysis resulted in more accurate results.

4 Discussion and Directions for Future Work

As it can be observed from presented results, predictions of stock prices depend strongly on choice of training dataset, their preparation methods and number of appearing messages per time interval. Predictions conducted with models trained with datasets with messages containing company stock symbol performs better. It can be explained by the fact that these messages refer to stock market. Tweets with company name may just transfer information, which does not affect financial results. Another important factor is a choice of preparation of training set. Two methods were used. One of the methods was a manual labelling sentiment value of messages. This method allows to more accurately label training data but is not effective for creating large training sets; at least with classical approach to manual labelling. The other method was applying SentiWordNet, which is a lexical resource for sentiment opinion mining. It enabled to create bigger training datasets, which resulted in building more accurate models. Last factor that is important for prediction is number of appearing messages per time interval. Although model trained with datasets with company name were not accurate in comparison to the other datasets, there is bigger number of tweets per time interval. It allowed performing prediction for shorter time intervals, which were not possible for dataset with messages containing company stock symbol. Described methods can be also used with other stock predictions procedures in order to maintain higher accuracy. It is also important to note that stock prediction methods are not able to predict sudden events called 'black swans' [16]. The last comment on the manual labelling in the context of latest development of crowdsourcing systems seems to be obvious. The future work for research in this subject is to convey a wide study of labelling large text datasets with aid of crowd intelligence as it is feasible and already conducted in other areas of research, especially concerning scientific images [17].

Predictions can be improved by adding analysis of metadata such as exact location of a person while posting message, number of retweets, number of followers etc. This information may be used to determine which users are more influential and creating a model of interactions between users. Number of messages posted by a user and its frequency may be used to discard spammers and automated Twitter accounts. It is also possible to employ different sources of information. Although Twitter is a popular social networking tool and offers nearly time communications there exist other sources of information such as different social networks, blogs, articles in online newspapers. Adding analysis of other may contribute to more accurate predictions.

5 Conclusions

This chapter discusses a possibility of making prediction of stock market basing on classification of data coming Twitter micro blogging platform. Results of prediction show that there is a correlation between information in social services and stock market. There are several factors that affect accuracy of stock predictions. First of all choice of datasets is very important. In the paper two types of datasets were used one with name of the company and the other with stock symbol. Predictions were made for Apple Inc. in order to ensure that sufficiently large datasets would be retrieved. There were large differences in size between the two sets of full company name and stock symbol mentioned within the twitter messages processed. This lead to situation that it was not possible to perform 5 min predictions based on tweets with stock symbol due to too few messages. Additionally although dataset with company name was bigger it may not be accurate for predictions. This time it may be effect of the fact that company name can be used as a household name and the messages may not refer to stock market. In case of tweets with stock symbol there is a greater probability that it was posted to be intentionally related to stock prices. Anyway, authors believe the presented case may be inspiring to develop complex algorithms especially building on the presented concept of computational environment and sentiment analysis has, in our opinion, significant role to play in research of economy trends as well as in general modelling of various aspects of life in coming years.

References

1. Y. Zhang, L. Wu, (2009). "Stock Market Prediction of S&P 500 via combination of improved BCO Approach and BP Neural Network". Expert systems with applications 36 (5): 8849–8854. doi:10.1016/j.eswa.2008.11.028
2. Z. Da, J. Engelberg, P. Gao: In Search of Attention, The Journal of Finance Volume 66, Issue 5, pages 1461–1499, October 2011, doi: 10.1111/j.1540-6261.2011.01679.x
3. J. Manyika, M. Chui, B. Brown, J. Bughin, R. Dobbs, C. Roxburgh, and A.H. Byers. Big data: The next frontier for innovation, competition, and productivity, McKinsey, May 2011.
4. Edd Dumbill. What is big data?: an introduction to the big data landscape. http://radar.oreilly.com/2012/01/what-is-big-data.html, 2012.
5. P. Zikopoulos, C.Eaton, D. DeRoos, T. Deutch and G. Lapis, Understanding Big Data: Analytics for Enterprise Class Hadoop and Streaming Data. McGraw-Hill Osborne Media, 2011
6. T. Ahlqvist and Valtion teknillinen tutkimuskeskus. Social media roadmaps: exploring the futures triggered by social media. VTT tiedotteita. VTT, 2008.
7. Twitter Statistics. http://www.statisticbrain.com/twitter-statistics/, 2016. [Online; accessed May-2016].
8. M. Paluch, L. Jackowska-Strumillo: The influence of using fractal analysis in hybrid MLP model for short-term forecast of closing prices on Warsaw Stock Exchange. Proceedings of the 2014 Federated Conference on Computer Science and Information Systems, M. Ganzha, L. Maciaszek, M. Paprzycki (eds). ACSIS, Vol. 2, pages 111–118 (2014) doi: 10.15439/2014F358
9. M. Marcellino, J. H. Stock, M.W. Watson, A comparison of direct and iterated multistep AR methods for forecasting macroeconomic time series, Journal of Econometrics Volume 135, Issues 1–2, November–December 2006, Pages 499–526 doi:10.1016/j.jeconom.2005.07.020
10. K-J. Kim, I. Han, Genetic algorithms approach to feature discretization in artificial neural networks for the prediction of stock price index, Expert Systems with Applications Volume 19, Issue 2, 2000, Pages 125–132 doi:10.1016/S0957-4174(00)00027-0
11. E. J. Ruiz, V. Hristidis, C. Castillo, and A. Gionis, "Correlating Financial Time Series with Micro-Blogging activity," WSDM 2012. Doi: 10.1145/2124295.2124358
12. B. Pang and L. Lee. Opinion mining and sentiment analysis. Found. Trends Inf. Retr.,2(1–2):1–135, January 2008, doi:10.1561/1500000011
13. Asur, S., Huberman, B.A., Predicting the Future with Social Media IEEE/WIC/ACM International Conference on Web Intelligence and Intelligent Agent Technology, 2010, pp 492–499 doi:10.1109/WI-IAT.2010.63
14. Y-W Seo, J.A. Giampapa, and K. Sycara. Text classification for intelligent portfolio management. Technical Report CMU-RI-TR 02-14, Robotics Institute, Pittsburgh, PA, May 2002.
15. A. Esuli and F. Sebastiani. Sentiwordnet: A publicly available lexical resource for opinion mining. In In Proceedings of the 5th Conference on Language Resources and Evaluation (LREC'06, pages 417–422, 2006. In In Proceedings of the 5th Conference on Language Resources and Evaluation (LREC'06, pages 417–422, 2006, doi: 10.1155/2015/715730

16. N.N. Taleb, Common Errors in the Interpretation of the Ideas of The Black Swan and Associated Papers (October 18, 2009)
17. C.Chen, P.Wozniak, A.Romanowski, M.Obaid, T.Jaworski, J.Kucharski, K.Grudzien, S.Zhao, M.Fjeld: Using Crowdsourcing for Scientific Analysis of Industrial Tomographic Images, ACM Transactions on Intelligent Systems and Technology (TIST), Vol. 7, Issue 4, May 2016, (in print).

On a Property of Phase Correlation and Possibilities to Reduce the Walsh Function System

Lubomyr Petryshyn and Tomasz Pełech-Pilichowski

Abstract Data processing algorithms are used for business, industry and public sector to filter input data, calculate values, detect abrupt changes, acquire information from data or to ensure signal consistency. It is an important research area for Big Data processing and processing of data received from Internet of Things. Typically, classical algorithms are exploited, i.e. statistical procedures, data mining techniques and computational intelligence algorithms. Referring to the area of signal processing, applications of mathematical transformation (e.g. Fourier Transform, Walsh–Fourier Transform) of input signals from either domain to the other are promising. They enable to perform complementary analyses and to consider additional signal components, in particular cyclic (periodic) ones (sin- and cos-components). The Walsh function system is a multiplicative group of Rademacher and Gray functions. In its structure, it contains discrete-harmonic, sin-components of the Rademacher functions, and cos-components of the Gray function, as well as discrete-irregular components of the Walsh function. In the paper, the phase interdependence property has been defined, in pairs of a complete Walsh function system. Odd (sin-components) and even (cos-components) Walsh function subsystems were extracted as theoretical and numerical processing databases. A perspective concerning the processing efficiency and digital signal processing is outlined.

1 Introduction

Information and communication technologies are widely exploited by business, industry and public sector require to provide reliable input data for processing with suitable numerical procedures. Dedicated algorithms for signal processing may be

L. Petryshyn · T. Pełech-Pilichowski (✉)
AGH University of Science and Technology, Al. Mickiewicza 30,
30-059 Krakow, Poland
e-mail: tomek@agh.edu.pl

L. Petryshyn
e-mail: petryshynL@poczta.fm

© Springer International Publishing AG 2017
T. Pełech-Pilichowski et al. (eds.), *Advances in Business ICT: New Ideas from Ongoing Research*, Studies in Computational Intelligence 658,
DOI 10.1007/978-3-319-47208-9_8

aimed at [1–3] time series analysis, forecasting/prediction, filtering, event/change detection [4], prediction or similarity analysis. Difficulties related to data processing include: data quality assessment issues, data preprocessing, data transformation and computational issues, which are essential for the processing of large data sets (including Big Data), high-performance computing, real-time processing [2] and data analyses for Internet of Things (IoT) purposes.

Datasets, including Big Data and information retrieved from heterogeneous sources (devices) typically are stored as time series. They may be considered as diagnostic signals. For further processing they need to be verified for consistency, completeness, authenticity and, finally, reliability. Moreover, such datasets may be redundant or noisy. This can cause difficulties in carrying out data analyses thus inabilities to use potentially valuable input data sets with available applications and systems. It also influences the usefulness of Business Intelligence solutions, business analytics, real-time reporting, probabilistic, statistical and predictive analyses, data mining techniques, and effectiveness of the use of learning algorithms. On the other hand, in many cases it is necessary to perform multiple input data (generally diagnostic signals) transformations for the data unification or standardization [2, 3], acquiring information from data, irrelevant information removal, emphasizing or deemphasizing specific attributes.

Interesting results of data processing (including Big Data) are obtained using data mining techniques [5], expert systems [6], neural networks [7] or evolutionary methods [8]. Nevertheless, research carried out in many Research Centers reveal a high potential of design and optimization of algorithms for digital signal processing (Fourier and Wavelet Transform, digital filtering etc.), for advanced data processing (especially for business purposes).

Referring to the area of signal processing, for data processing purposes applications of mathematical transformations (e.g. Fourier Transform, Walsh–Fourier transform) of input signals from either domain to the other are promising. They enables to perform complementary analyses and to consider additional signal components, in particular cyclic (periodic) ones sinus and cosines components. On the other hand, advanced procedures (for example, statistical or digital-signal-processing-based) allow to obtain robust results but they usually are time-consuming.

Digital signal processing may require utilization of high computational power, especially for Big Data processing. Thus, the effectiveness of implementation of signal processing algorithms should be considered as a key factor of its technical and economic evaluation [9–11].

Historically, the basic functions were mostly used as harmonic and discrete-harmonic *sin*- and *cos*-like functions, due to the simplicity of technical solutions with respect to the implementation of qualitative and stable generators of harmonic functions systems in digital signal processing [12–17]. The development of digital technology also enabled an easy implementation of other system functions, including discrete-non-harmonic Walsh and Galois functions [11, 18–21], which have an extended functionality compared to discrete-harmonic function systems [20–23]. A possibility of technical implementation led to a rapid development of mathematical, theoretical and numerical processing, on the basis of which systems were developed, including digital formation, transmission, filtering, processing, compression and storage of data with expanded or improved technical and economic parameters [10, 11, 18–23].

The nature of an information source determines the properties of the formed information stream, which can be further analyzed and synthesized using only even functions, only odd ones or both types [10, 12, 21, 22]. On the other hand, the range of the theoretical and numerical processing may be artificially matched with the nature of only even or odd function systems [11, 22, 23]. Thus, when digital processing of input stream data is performed, only even function systems may be used (e.g. the Gray function system) or odd ones (e.g., the Rademacher function system). In this case it necessary to fix the phase parameter of component functions, or one can use a system containing both even and odd function components (e.g., a discrete-phase Walsh Galois ones, etc.) without having to fix any zero input signal phases, which leads a reduction in the number of operations in the processing of an input signal zero phase. However, there is a problem to determine the optimal computing power to process a larger/smaller number of basic function system components, and the related need to compute, resulting from the presetting of the input data stream initial phase.

Initially, studies showed that the Walsh function complete system contains the same pairs of odd and even functions, with cross-phase shift of $\pm\pi/2$, in the period of component pair definition for each function order. Such a feature allows to use of a function system to carry out theoretical and numerical processing, as well as digital signal processing in a full set of odd and even components, and additionally in two partial options—a system containing only even or odd functions. To determine the direction of further research, we will examine how Walsh functions are formed and what their cross-phase interactions are.

2 Discrete-Harmonic Function Systems

With theoretical and numerical processing based on functional analysis and synthesis, one should consider a number of parameters, of which, according to the problem formulated in the Introduction, we shall analyze a number of functions in a basic Walsh set, and their nature, concerning their evenness or odd parity. In the field of digital signal processing, one can eliminate of some operations due to the nature of the signal which is a subject of an analysis and synthesis. In particular, we shall examine a structural specificity of basis function systems that are even and odd, for theoretical and numerical processing, which requires the necessity and sufficiency to use systems of even and odd functions, without considering phase shifts.

One of classical functional analysis bases is the Rademacher base [9–14], which is characterized by a discrete, orthonormal function system (Eq. 1).

$$Rad(n, \theta) = \text{sign}[sin\ 2n\pi, \theta] \tag{1}$$

where $i = 0, 1, 2, \ldots, n$ denotes a function order number, θ—time parameter, time-normalized to a period of $T : \theta = t/T$, t denotes current time value, $n = log_2 N$ and describes the order of basic function system referring to theoretical and numerical processing, N denotes the integer module in the function system.

$$signx\,(t) = \begin{cases} 1 & if\ x\,(t) > 0 \\ -1 & if\ x\,(t) < 0 \ , \\ 0 & if\ x\,(t) = 0 \end{cases} \tag{2}$$

Formula (1) draws our attention to the fact that the Rademacher function is considered as "pure" *sin*-components, which change their value at point $t = 0$. In other words, the Rademacher base is a system of odd functions characterized by the following relation (Eq. 3):

$$F(-t) = -F(t) \tag{3}$$

With this arrangement, without additional phase shifts, one can analyze and synthesize an odd function.

On the other hand, a Gray function system is known [20, 21, 23]. It is formed in the following dependency (Eq. 4):

$$Gry(n, \theta) = sign[cos\ 2^n \pi, \theta] \tag{4}$$

Unlike with the Rademacher base, the Gray base is a system of functions so-called "clean" *cos*-components, which have a nature of even functions, i.e. (Eq. 5):

$$F(-t) = F(t) \tag{5}$$

Therefore, using the Gray function, without additional phase shifts, one can analyze even type functions. However, if an analysis of even and odd functions must be carried out, it is necessary to choose one of the following solutions.

It is known that a phase shift interdependence at a value of $\pm\pi/2$ is found for odd or sin- components, or even and cos-components. Thus, the first decision is to choose one of any of even or odd function systems assuming that there is a phase shift.

Another option is to choose one combined system of even and odd functions, without having to establish their phase zero by a value of $\pm\pi/2$. It was originally proposed to use a basic discrete function system [23], following its combination of Gray even function system, and Rademacher odd functions, as follows (Eq. 6):

$$DF(n, \theta, i) = \begin{cases} Rad\ (n, \theta) = sign[sin\ 2^n \pi, \theta] \\ Gry\ (n, \theta) = sign[cos\ 2^n \pi, \theta] \end{cases} \tag{6}$$

Figure 1 shows an example of functions of the order between 0 and 3.

Fig. 1 Sample system of Gray even function system, and Rademacher odd functions

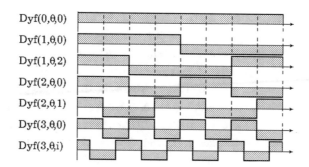

Dyf(0,θ,0)

Dyf(1,θ,0)

Dyf(1,θ,2)

Dyf(2,θ,0)

Dyf(2,θ,1)

Dyf(3,θ,0)

Dyf(3,θ,i)

Effectiveness studies on the solutions presented above are not the main subject of this paper while it describes a possibility of further research (the research results will be published).

Analyses of classical base systems of Rademacher and Gray functions and the proposed function system enable to analyze functional properties and the application efficiency of the Walsh function.

3 Synthesis of the Walsh Function

The basis for the Walsh function was developed in 1923 [15, 19, 20], and was widely applied, due to the expansion of the system functionality [19–21, 23]. Function systems which are arranged according to Walsh, Paley, Hadamard, or otherwise, contain a complete invariant set of functions [20, 21, 23], or a full, multiplicative group of Rademacher and Gray [23] function components. Below, a basic structure of a discrete trigonometric Walsh functions $Wal(n, \theta, i)$, arranged according to Walsh, on the basis of Rademacher $Rad(n, \theta)$ and Gray $Gry(n, \theta)$ functions is presented.

$$Wal(0, \theta) = Rad(0, \theta) = sign[sin\pi] = Gry(-1, \theta) = sign[cos\pi/2],$$

$$Wal(1, \theta, 1) = Rad(1, \theta) = sign[sin2\pi] = Gry(0, \theta) = sign[cos\pi],$$

$$Wal(1, \theta, 2) = Rad(1, \theta) Rad(2, \theta) = sign[sin2\pi] sign[sin4\pi] = Gry(1, \theta) = sign[cos2\pi],$$

$$Wal(2, \theta, 1) = Rad(2, \theta) = sign[sin4\pi] = Gry(0, \theta) Gry(1, \theta) = sign[cos\pi] sign[cos2\pi],$$

$$Wal(2, \theta, 2) = Rad(2, \theta) Rad(3, \theta) = sign[sin4\pi] sign[sin8\pi] = Gry(2, \theta) = sign[cos4\pi],$$

$$Wal(3, \theta, 1) = Rad(1, \theta) Rad(2, \theta) Rad(3, \theta) = sign[sin2\pi] sign[sin4\pi]sign[sin8\pi] =$$
$$= Gry(0, \theta) Gry(2, \theta) = sign[cos\pi] sign[cos4\pi],$$

$$Wal(3, \theta, 2) = Rad(1, \theta) Rad(3, \theta) = sign[sin2\pi] sign[sin8\pi] = Gry(1, \theta) Gry(2, \theta) =$$
$$= sign[cos2\pi] sign[cos4\pi],$$

$$Wal(3, \theta, 3) = Rad(3, \theta) = sign[sin8\pi] = Gry(0, \theta) Gry(1, \theta) Gry(2, \theta) =$$
$$= sign[cos\pi]sign[cos2\pi] sign[cos4\pi],$$

$$Wal(3, \theta, 4) = Rad(3, \theta) Rad(4, \theta) = sign[sin8\pi] sign[sin16\pi] = Gry(3, \theta) = sign[cos8\pi],$$

$$Wal(4, \theta, 1) = Rad(1, \theta)\,Rad(3, \theta)Rad(4, \theta) = sign[sin2\pi]\,sign[sin8\pi]\,sign[sin16\pi] =$$
$$= Gry(0, \theta)\,Gry(3, \theta) = sign[cos\pi]\,sign[cos8\pi],$$

$$Wal(4, \theta, 2) = Rad(1, \theta)\,Rad(2, \theta)\,Rad(3, \theta)\,Rad(4, \theta) =$$
$$= sign[sin2\pi]\,sign[sin4\pi]\,sign[sin8\pi]\,sign[sin16\pi] =$$
$$= Gry(1, \theta)\,Gry(3, \theta) = sign[cos2\pi]\,sign[cos8\pi],$$

$$Wal(4, \theta, 3) = Rad(2, \theta)\,Rad(3, \theta)\,Rad(4, \theta) = sign[sin4\pi]\,sign[sin8\pi]\,sign[sin16\pi] =$$
$$= Gry(0, \theta)\,Gry(1, \theta)\,Gry(3, \theta) = sign[cos\pi]\,sign[cos2\pi]\,sign[cos8\pi],$$

$$Wal(4, \theta, 4) = Rad(2, \theta)\,Rad(4, \theta) = sign[sin4\pi]\,sign[sin16\pi] =$$
$$= Gry(2, \theta)\,Gry(3, \theta) = sign[cos4\pi]\,sign[cos8\pi],$$

$$Wal(4, \theta, 5) = Rad(1, \theta)\,Rad(2, \theta)\,Rad(4, \theta) = sign[sin2\pi]\,sign[sin4\pi]\,sign[sin16\pi] =$$
$$= Gry(0, \theta)\,Gry(2, \theta)\,Gry(3, \theta) = sign[cos\pi]\,sign[cos4\pi]\,sign[cos8\pi].$$

$$Wal(4, \theta, 6) = Rad(1, \theta)\,Rad(4, \theta) = sign[sin2\pi]\,sign[sin16\pi]$$
$$= Gry(1, \theta)\,Gry(2, \theta)\,Gry(3, \theta) == sign[cos2\pi]\,sign[cos4\pi]\,sign[cos8\pi],$$

$$Wal(4, \theta, 7) = Rad(4, \theta) = sign[sin16\pi] = Gry(0, \theta)\,Gry(1, \theta)\,Gry(2, \theta)\,Gry(3, \theta) =$$
$$= sign[cos\pi]\,sign[cos2\pi]\,sign[cos4\pi]\,sign[cos8\pi]$$

Fig. 2 Walsh functions arranged according to Walsh

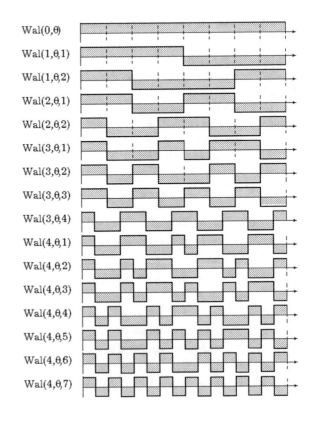

Wal(0,θ)
Wal(1,θ,1)
Wal(1,θ,2)
Wal(2,θ,1)
Wal(2,θ,2)
Wal(3,θ,1)
Wal(3,θ,2)
Wal(3,θ,3)
Wal(3,θ,4)
Wal(4,θ,1)
Wal(4,θ,2)
Wal(4,θ,3)
Wal(4,θ,4)
Wal(4,θ,5)
Wal(4,θ,6)
Wal(4,θ,7)

Walsh functions arranged according to Walsh, or according to frequency (see Fig. 2) include discrete harmonic Rademacher and Gray functions, and the original set of discrete-irregular, multiplied Walsh functions $Wal(n, \theta, i)$.

4 Phase Correlation Properties of the Walsh Function System. Even and Odd Walsh Function Sub-systems

If Rademacher and Gray functions in the Walsh base system form complete systems of discrete-harmonic functions, also other selected discrete-irregular functions will create, respectively, even and odd components of the function system. For example, for a fourth-order Walsh function system (see Fig. 3) the following phase interdependency was defined.

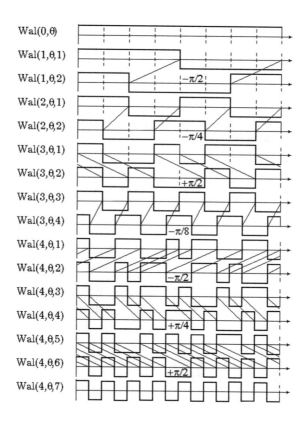

Fig. 3 The phase interdependency of a Walsh function system

$$Wal(0, \theta) = sign[sin\pi] = sign[cos\pi/2]$$
$$Wal(1, \theta, 1) = Wal(1, \theta, 2) + \pi/2,$$
$$Wal(1, \theta, 2) = Wal(1, \theta, 1) - \pi/2,$$
$$Wal(2, \theta, 1) = Wal(2, \theta, 2) + \pi/4,$$
$$Wal(2, \theta, 2) = Wal(2, \theta, 1) - \pi/4,$$
$$Wal(3, \theta, 1) = Wal(3, \theta, 2) - \pi/2,$$
$$Wal(3, \theta, 2) = Wal(3, \theta, 1) + \pi/2,$$
$$Wal(3, \theta, 3) = Wal(3, \theta, 4) + \pi/8,$$
$$Wal(3, \theta, 4) = Wal(3, \theta, 3) - \pi/8,$$
$$Wal(4, \theta, 1) = Wal(4, \theta, 2) + \pi/2,$$
$$Wal(4, \theta, 2) = Wal(4, \theta, 1) - \pi/2,$$
$$Wal(4, \theta, 3) = Wal(4, \theta, 4) - \pi/4,$$
$$Wal(4, \theta, 4) = Wal(4, \theta, 3) + \pi/4,$$
$$Wal(4, \theta, 5) = Wal(4, \theta, 6) - \pi/2,$$
$$Wal(4, \theta, 6) = Wal(4, \theta, 5) + \pi/2,$$
$$Wal(4, \theta, 7) = Wal(5, \theta, 1) + \pi/16,$$
$$Wal(5, \theta, 1) = Wal(4, \theta, 7) - \pi/16.$$

As outlined above, due to sin-component extraction, we can obtain a subsystem of odd Walsh functions.

$$Wal(0, \theta) = Rad(0, \theta),$$
$$Wal(1, \theta, 1) = Rad(1, \theta),$$
$$Wal(2, \theta, 1) = Rad(2, \theta),$$
$$Wal(3, \theta, 1) = Rad(1, \theta) \, Rad(2, \theta) \, Rad(3, \theta),$$
$$Wal(3, \theta, 3) = Rad(3, \theta),$$
$$Wal(4, \theta, 1) = Rad(1, \theta) \, Rad(3, \theta) \, Rad(4, \theta),$$
$$Wal(4, \theta, 3) = Rad(2, \theta) \, Rad(3, \theta) \, Rad(4, \theta),$$
$$Wal(4, \theta, 5) = Rad(1, \theta) \, Rad(2, \theta) \, Rad(4, \theta),$$
$$Wal(4, \theta, 7) = Rad(4, \theta)$$

Cosine-components of the Walsh function system, as a subsystem of even functions are extracted below.

$$Wal(0, \theta, 1) = Gry(-1, \theta),$$
$$Wal(1, \theta, 1) = Gry(0, \theta),$$
$$Wal(1, \theta, 2) = Gry(1, \theta),$$
$$Wal(2, \theta, 2) = Gry(2, \theta),$$
$$Wal(3, \theta, 2) = Gry(1, \theta)\, Gry(2, \theta),$$
$$Wal(3, \theta, 4) = Gry(3, \theta),$$
$$Wal(4, \theta, 2) = Gry(1, \theta)\, Gry(3, \theta),$$
$$Wal(4, \theta, 4) = Gry(2, \theta)\, Gry(3, \theta),$$
$$Wal(4, \theta, 6) = Gry(1, \theta)\, Gry(2, \theta)\, Gry(3, \theta),$$
$$Wal(5, \theta, 1) = Gry(4, \theta).$$

A complete n-order Walsh function basic system with a functional analysis guarantees a minimal discretization step within a time scale of $\Delta\theta_n = \pi/2^{n-1}$, with all values of $n\Delta\theta_n = n\pi/2^{n-1}$ for respective $n = n, n-1, \ldots, 1, 0$, orders, and a corresponding, discrete supplementation of function system: $\Delta\theta_i = \pi/2^{i-1}$, for all n, where $i = n-1, \ldots, 1, 0$.

As a result of the extraction of those very *sin*-and *cos*-components from a complete Walsh function system, we can separate Rademacher and Gray function systems, from which it can be stated that the Walsh function system contains complete discrete-harmonic Rademacher and Gray bases, as well as a sub-system of discrete-irregular functions, which is a complete multiplicative system of Rademacher or Gray functions.

Based on the presented above results, it can be stated that the Walsh function base is an advanced function system that allows an effective implementation of a functional analysis. However, the Walsh system is characterized by a relatively high functional redundancy due to the fact that the computational power (complexity) the function base is $P = N^2$ [24].

Presently, more efficient systems are known as the Galois function [20, 21, 23]. The transition and the creation of its basic function systems is carried out by a theoretical and numerical transformation of the Walsh base. Considerations outlined above draw a conclusion on further research of the authors. After the analogy of Walsh function systems, arranged by well-known principles of by Walsh, Paley, and Hadamard [14, 15, 19, 20, 23], a new, original principle can be introduced. Recursive ordering of the Walsh function, on whose basis several new basic Walsh function systems will be created, which in turn are the basis for the synthesis of Galois function systems.

5   Conclusions

Reliable analyses require the use of varied methods, approaches or paradigms to enable further processing or to obtain useful information. On the other hand, despite the increasing computing power, it's reasonable to implement fast algorithms which allow for performing analysis in real-time. It is an important factor for business purposes where the speed of obtaining a result may be crucial (e.g. financial time-series processing, supervisor control, systems of state critical architecture). In the paper a subset of advanced algorithms based on mathematical transformation were analyzed which may be used for further data processing (time series, Big Data, IoT data resources etc.).

It was discussed that the Walsh function system is a complete, multiplicative group of Rademacher and Gray functions. The property of phase correlation of function components in pairs of the same order was established. It allows their grouping in sub-systems of even type (*sin*-components) and odd (*cos*-component) functions. In systems of each type, we defined discrete-harmonic and discrete-irregular sub-functions. Each of even or odd type sub-functions can be used in signal digital processing as a system of basic functions. The study was of an analytical kind. It outlined new properties of the phase correlation in the Walsh function system (and Gray/Rademacher functions). It determines a direction of further research on the effectiveness in the application of these systems in functional analysis of IT processes, time series processing and digital signal processing.

References

1. Duda J.T., Pełech-Pilichowski T., Enhancements of moving trend based filters aimed at time series prediction. [In:] Advances in systems science, Eds. Świągonatek J. et al., Springer, 2014
2. Pełech-Pilichowski T., Non-stationarity detection in time series with dedicated distance-based algorithm. [In:] Frontiers in information technology, Ed. Al-Dahoud A., Masaum Net., 2011
3. Pełech-Pilichowski T., Duda J.T., A two-level detector of short-term unique changes in time series based on a similarity method. Expert Systems, Wiley (early view; to be printed)
4. Pełech-Pilichowski T., Duda J.T.: A two-level detector of short-term unique changes in time series based on a similarity method. Expert Systems, Vol. 32 Issue 4, August 2015, Elsevier, pp. 555–561
5. Shmueli G., Patel N.R., Bruce P.C., Data Mining for Business Intelligence: Concepts, Techniques, and Applications in Microsoft Office Excelwith XLMiner, 2nd Edition, Wiley 2010
6. Krishna T.G., Abdelhadi M.A., Expert Systems in Real world Business. International Journal of Advance Research in Computer Science and Management Studies, Vol.1., Issue 7, International Journal of Advance Research in Computer Science and Management Studies 2013
7. He X., Xu S., Process Neural Networks, Springer (Jointly published with Zhejiang University Press), 2010
8. Ferreira T.A.E., Vasconcelos G.C., Adeodato, P. J. L., A new evolutionary method for time series forecasting. [In:] ACM Proceedings of Genetic Evolutionary Computation Conference-GECCO 2005, Washington, DC. ACM Publ.
9. J. G. Proakis, Digital Signal Processing: Principles, Algorithms, and Applications, Pearson Education, 2007, p. 1156.

10. S. W. Smith, Digital Signal Processing: A Practical Guide for Engineers and Scientist, Newnes, 2003, p. 650.
11. R. E. Blahut, Fast algorithms for digital signal processing, Addison-Wesley Pub. Co., 1985, p. 441.
12. M. G. Karpovskii, E. S. Moskalev, Spektral'nye metody analiza i sinteza diskretnyh ustroistv. -L., Energiya, 1973, p. 144, (in Russian).
13. H. Rademacher, Einige Satze von allgemeine Ortogonalfunktionen, Math. Annalen, 1922, N 87, pp. 122–138.
14. R. E. A. C. Paley, A Remarkable Series of Ortogonal Funktions, Proc. London Math. Soc., 1932, (2)34, pp. 241–279.
15. J. L. Walsh, A closed set of ortogonal functions, Amer. J. of Mathematics, 1923, V.45, pp. 5-24.
16. A. Haar, Zur Theorie der ortogonalen Funktionsysteme, Math. Ann., 1910. V.69. pp. 331–371; 1912, V.71. pp. 38–53.
17. B. Gold, C. M. Rader, Digital processing of signals, McGraw-Hill, 1969, p. 269.
18. A. V. Oppenheim, Discrete-Time Signal Processing, Pearson Education, 2006, p. 864.
19. B. I. Golubov, A. V. Efimov, V. A. Skvorcov, Ryady i preobrazovaniya Walsh'a: Teoriya i primeneniya, Nauka, 1987, p. 343, (in Russian).
20. L. A. Zalmanzon, Preobrazovaniya Fourier'a, Walsh'a, Haar'a i ih primenenie v upravlenii, svyazi i drugih oblastyah, -M., Nauka, 1989, p. 496, (in Russian).
21. L. V. Varichenko, V. G. Labunec, M. A. Rakov, Abstraktnye algebraicheskie sistemy i cifrovaya obrabotka signalov, -Kiev, Naukova Dumka, 1986, p. 248, (in Russian).
22. L. R. Rabiner, B. Gold, Theory and Application of Digital Signal Processing, Prentice Hall, 1975, p. 762.
23. L. B. Petryshyn, Teoretychni osnovy peretvorennya formy ta cyfrovoi obrobky informacii v bazysi Galois'a. –Kyiv, IZiMN MOU, 1997, p. 237, (in Ukrainian).
24. V. M. Mutter, Osnovy pomehoustoichivoi teleperedachi informacii, M. Energoatomizdat, 1990, p. 288, (in Russian).

Printed in the United States
By Bookmasters